Engineering Materials

This series provides topical information on innovative, structural and functional materials and composites with applications in optical, electrical, mechanical, civil, aeronautical, medical, bio- and nano-engineering. The individual volumes are complete, comprehensive monographs covering the structure, properties, manufacturing process and applications of these materials. This multidisciplinary series is devoted to professionals, students and all those interested in the latest developments in the Materials Science field, that look for a carefully selected collection of high quality review articles on their respective field of expertise.

Indexed at Compendex (2021)

Min Jiang · Xinhua Wang

Slag-Steel Reaction and Control of Inclusions in Al Deoxidized Special Steel

Min Jiang
School of Metallurgical and Ecological Engineering
University of Science and Technology Beijing
Beijing, China

Xinhua Wang
School of Metallurgical and Ecological Engineering
University of Science and Technology Beijing
Beijing, China

ISSN 1612-1317 ISSN 1868-1212 (electronic)
Engineering Materials
ISBN 978-981-19-3465-0 ISBN 978-981-19-3463-6 (eBook)
https://doi.org/10.1007/978-981-19-3463-6

Jointly published with Metallurgical Industry Press, Beijing, China
The print edition is not for sale in China (Mainland). Customers from China (Mainland) please order the print book from: Metallurgical Industry Press.

This Springer imprint is published by the registered company Springer Nature Singapore Pte Ltd.
The registered company address is: 152 Beach Road, #21-01/04 Gateway East, Singapore 189721, Singapore

Preface

Control of non-metallic inclusions has been a very hot topic in recent decades. Steelmakers always make efforts for cleaner steel with decreased amounts of inclusions for improved performances of steel products. Unfortunately, formation of inclusions is inevitable because of intensive using of oxygen in primary steelmaking. As a result, natures of inclusions including the number, size, chemistry, etc. should be strictly controlled.

Al deoxidation is widely used in steelmaking for its strong affinity to dissolved oxygen in liquid steel. Combined with basic slag refining, higher cleanliness can be achieved in steel. However, the produced inclusions were usually featured by high melting points. They often cause nozzle clogging problems in continuous casting and also are complained for various defects in final products during service. Because of that, such inclusions often deteriorate mechanical properties of steel. Hence, how to target inclusions with improved deformability, viz. with low melting temperature is of great significance to Al deoxidized steels.

In present book, control of inclusions in Al deoxidized steel is intensively discussed. The authors want to share with the reader important studies by us in recent years, that is, targeting low melting point inclusions in Al killed special steels. Based on systematical laboratorial study on slag-steel reactions and its influence on the control of inclusions, the authors proposed the high-basicity calcium aluminate refining slag, which is very favorable to target liquid/semi-liquid inclusions. The evolution phenomenon of inclusions from alumina to spinel and finally to low melting point calcium magnesia aluminate were described, and the mechanism was elucidated. Exploring study was also carried out to clarify the influence of refractory materials on the evolution of inclusions. Moreover, based on the understanding of chemical reactions among slag, steel and inclusions, prohibiting the formation of CaO-containing inclusions in steel was also explored.

Beijing, China

Min Jiang
Xinhua Wang

Acknowledgements The authors want to express sincere appreciation and gratitude to the support of State Key Laboratory of Advanced Metallurgy, University of Science and Technology Beijing for the financial support to the publication of this book.

Appreciations are also expressed to graduate students of our team (Mr. Li Kai-lun, Wang Rui-gang, Yao Tong, Liu Ji-chen, Wang Zhang-yin, Zhong Hua-jun, Wang Kang-hao, Liu Jia-ming and Zhang Pei) for helping with edition.

Contents

Chapter 1
A Brief Review of Inclusions in Al Deoxidized Steel

Abstract Control of inclusions is a very hot topic in recent decades in steelmaking field because of the important influences of inclusions on the production, fabrication and performances of steels. However, inclusions cannot be avoided because steelmaking process is featured by intensive using of oxygen. As a result, how to remove inclusions from steel as much as possible is always a crucial task of steelmakers. However, inclusions cannot be completely removed from steel. So, natures of the remained inclusions must be strictly controlled, including the sizes, chemistry, morphologies and so on. At present, commercial steel grades can be roughly divided into three types by the choices of deoxidation in steelmaking. Most of the commercial steel grades is either with Al deoxidation or Si deoxidation, while only a few of them is with other deoxidation method such as Si–Al complex deoxidation. Because of wide application of Al as the deoxidant, how to control inclusions in Al-deoxidized steels is significant in both research and industrial production. In this chapter, control of inclusions is briefly reviewed and emphasis is put to Al deoxidize steels.

In recent decades, great efforts have been made in steel industry for ever enhanced strength, toughness and anti-fatigue property of steel materials for sustainable development of our society. In the late 1990s, Japan launched the research project for a new generation of steel material, which was also called the STX-21 project for super steel. The goal was to double the tensile strength and serving life of structural steels which were widely used in the constructions of infrastructures such as road, bridges, high building and so on. Afterwards, Chinese government started the key research project for the new generation of steel material, and the goal was to double the tensile strength and fatigue life of specific steel products and the goal has been realized after great efforts in research and industrial production [1, 2].

High strength low alloy steel such as spring steel, gearing steel, bearing steel and so on were usually used as important parts in manufacturing industry. For example, springs have important applications in both autos and trains. Because of periodical and dynamic loads in service, fatigue problems often occurred to them, and it was found that fatigue problems were often caused by inclusions in steel. Because of important influences of inclusions on the performances of steels, great attentions

M. Jiang and X. Wang, *Slag-Steel Reaction and Control of Inclusions in Al Deoxidized Special Steel*, Engineering Materials, https://doi.org/10.1007/978-981-19-3463-6_1

have been paid to the control of inclusions in steelmaking process in recent decades in China.

1.1 Classification of Inclusions

During steelmaking, chemical reactions between metallic elements including [Al], [Si], [Ca] etc. and non-metallic elements like [O], [S], [N] would occur, resulting in the formation of various non-metallic compounds. Such non-metallic compounds are ordinarily classified by their chemical compositions, deformability and origins.

According to chemical compositions, inclusions can be classified as sulfide, oxide, nitride and carbon nitride and so on. Because of intensive using of oxygen in steelmaking, oxide inclusions aroused great attentions of steelmakers. In industrial practice, oxide inclusions existing in steel can be varied, such as Al_2O_3, spinel, calcium aluminates, and so on. During the industrial production, inclusions in rolling products are often inspected by the optical microscopy and rated according to the ASTM E45-05 standard of inclusions. According to the standard, inclusions are usually categorized into the A-type, B-type, C-type and D-type by their morphologies along rolling direction. In China, the standard (GB/T 10,561-2005) to evaluate inclusions is originated from the ASTM standard. Moreover, in this standard, when the D-type inclusions are bigger than 12 μm, they would be named as the DS-type inclusions [3, 4].

In industrial production, inclusions in final products can also be divided by their deformability in hot rolling [5]. According to the deformability, inclusions are usually categorized as:

(1) Plastic inclusions. It means that the inclusion can be well deformed into thin strips along hot rolling direction. Such plastic inclusions ordinarily include FeS, MnS, (Mn,Fe)S and silicates with SiO_2 about 40–60%.
(2) Fragile inclusions. It means that the inclusion would be crushed into a chain of smaller pieces along the rolling direction. The inclusions mainly include Al_2O_3 and nitrides of V, Ti, Zr etc.
(3) Un-deformable inclusions. It means that the inclusions can hardly be deformed and would keep their original shapes. Hence, they exist as singular particles after hot rolling. Such inclusions are usually with high melting points.

Behaviors of inclusions in the rolling process would greatly affect their natures in steel and thus indicate different influences on the performances of final products.

Malkiewicz and Rudnik proposed and defined the deformation index, which can be expressed as $\upsilon = E_B/E_M$, to evaluate the deformation behaviors and abilities of inclusions [6]. Where, E_B and E_M are the real deformation of inclusions and steel matrix, respectively. For example, if a spherical particle with diameter of a was transformed into an oval shaped particle, then its deformation behavior in the rolling process can be expressed as following:

$$\upsilon = 2/3(\lg \lambda / \lg H) \tag{1.1}$$

where, λ is the ratio of the major axis and minor axis of the ellipse, viz., b/a, which is used to evaluate the deformation of the inclusion. While H is the ratio of the thickness of steel sample before and after rolling, viz. h_1/h_2, which can be used to characterize the deformation of steel matrix.

According to Rudnik, if the deformation index of inclusions were in the range of 0.5–1.0, micro cracks were seldom initiated by inclusions. By comparison, when the deformation index of inclusions was in the scope of 0.03–0.5, micro cracks can be caused at the steel-inclusion interfaces. When the deformation index of inclusions were furtherly decreased to less than 0.03, voids can be formed at steel-inclusion interfaces or tears would be even resulted at the steel-inclusion interfaces. In Fig. 1.1, formation of voids and micro cracks around inclusions during the rolling of steel were schematically depicted [7, 8].

Kiessling compared the deformation behaviors of various types of inclusions, including (Fe,Mn)O, Al_2O_3, calcium aluminates, spinel ($AO{\cdot}B_2O_3$) inclusions, silicates and sulfides to deduce and summarize the influences of rolling temperature on the deformability of inclusions. It was found that Al_2O_3, calcium aluminates and spinel etc. can hardly be deformed in hot rolling. By contrast, MnS can be well elongated into strip and indicate nearly identical deformation to steel at about 1000 °C [9].

It has been found that deformability of inclusions is greatly affected by their chemical compositions, which in turn influence the melting temperatures of them. For inclusions composed of $MnO–Al_2O_3–SiO_2$ system, when their compositions located inside or neighbor to the spessartine ($3MnO{\cdot}Al_2O_3{\cdot}3SiO_2$) region that with low melting temperature, they can be well deformed into strips during the hot rolling.

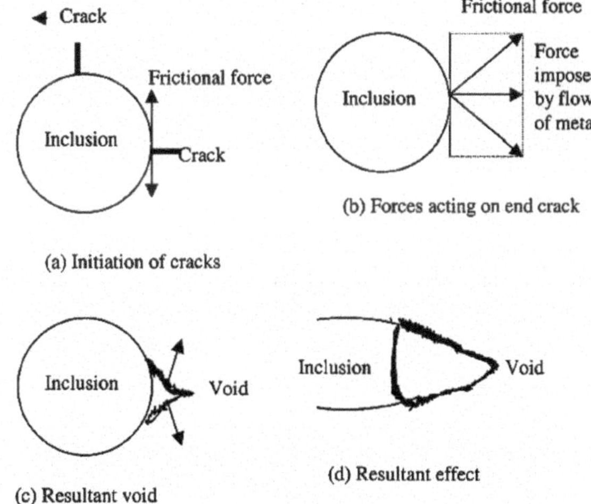

Fig. 1.1 Formation of voids and micro cracks around inclusions during rolling [8], reprinted from Progress in Materials Science, vol. 48, Atkinson HV and Shi G, Characterization of inclusions in the steel: a review including the statistics of extreme method progress in material, 457–520, copyright 2002, with permission from Elsevier

For inclusions composed of $CaO–Al_2O_3–SiO_2$ system, when their chemical compositions locate inside or next to the low melting point region surrounded by anorthite ($CaO·Al_2O_3·2SiO_2$), pseudo-wollastonite and Gehlenite, they would also show good deformability in the hot rolling of steel [10].

1.2 Fatigue Defects Initiated by Inclusions

Fatigue breakage is a typical type of problem that can be often observed during the service of steel materials. Fatigue defects of steel materials are detrimental in engineering, which aroused great attentions of engineers and had been intensively studied for many years. It is found that fatigue failures of steel (over 10^7) are often caused by inclusions, because inclusions can easily initiate stress concentration at steel-inclusion interfaces to cause micro cracks and voids. Hence, decreased amounts of inclusions would be more desirable.

Many studies proved that higher cleanliness of steel, viz., decreased impurities and inclusions in steel, would be very helpful to enhance anti-fatigue property of steel. Moreover, inclusions with poor deformability can much easier initiate micro cracks in steel matrix. Besides, the negative influence of inclusions to steel also relates to their chemical compositions. For example, TiN can be more detrimental than Al_2O_3 and spinel with same sizes. By contrast, calcium aluminates and CaS of the same size would be less negative. Whereas, in steel with Ca-treatment, large calcium aluminates are often formed, which can also be very dangerous. Consequently, Ca-treatment is denied in the steelmaking of bearing steel [11, 12].

Theoretical calculation revealed that decreasing melting temperature would be quite helpful to improve the deformability of inclusions thus to reduce stress concentration caused by inclusions [13]. According to Bernard [14], when inclusions can be softened at a temperature lower than 1623 K, they would be well deformed in the hot rolling of steel.

To explain the effects of inclusions on fatigue resistance property of steels, Brooksbank and Andrews proposed and established the B-A model, which emphasized the influences of thermal expansion of inclusions [15–18]. During the heating and cooling process of steels, because of distinctive thermal expansion of steel and inclusions, stress would be initiated at the interface of inclusion and steel. And the value of the stress closely relates to the rolling temperature and expansion coefficients of inclusions as well as the steel. This model can well explain why the negative effects of fragile oxides can be well relived when they were surrounded by MnS, and can also explain why large calcium aluminates were more detrimental than Al_2O_3 in bearing steel. Brooksbank and Andrews depicted the potential of various kinds of inclusions in initiating stress concentrations and voids in steel [18], from which Fig. 1.2 can be redrawn to compare the negative effects of typical inclusions in Al deoxidized steels.

Kiessling and Nordberg reported that influences of inclusions on the fatigue resistance of steel were mainly attributed to their deformability in hot rolling of steel. Hence, deformability of various kinds of inclusions were compared and they

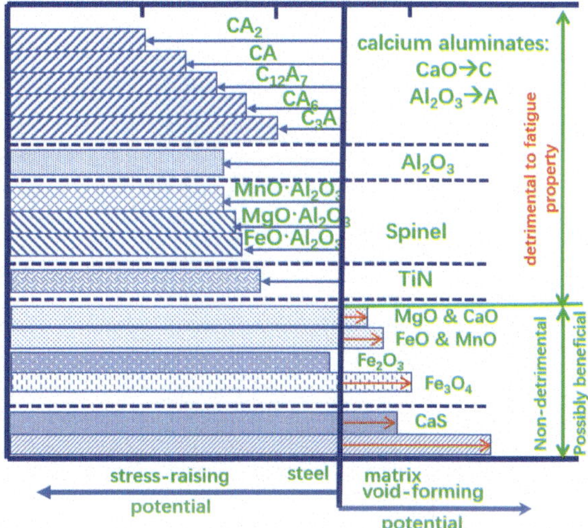

Fig. 1.2 Average thermal expansion coefficients of typical inclusions in steel

proposed the K-N model to explain the influences of inclusions with different deformability [19–21]. According to their model, inclusions having good deformability can maintain good contact to steel matrix without formation of voids at steel-inclusion interfaces. From the results in the studies, Fig. 1.3 can be redrawn to show typical inclusions in Al and Si deoxidized steels. Usually, inclusions would be more dangerous with the rise of their sizes [22]. Moreover, according to Zhang, steels with higher strength are more sensitive to inclusions [23].

To evaluate negative influences of inclusions' sizes on the fatigue resistance of steel, Duckworth proposed the concept of "critical size" for inclusions [24]. It

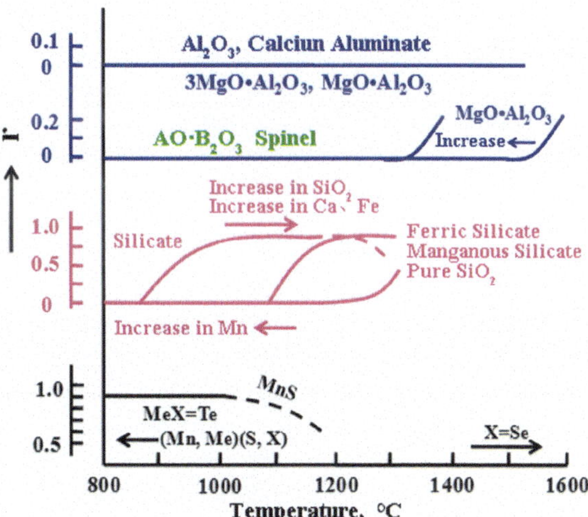

Fig. 1.3 Deformability of typical inclusions in steel

is proposed that when the inclusions located beneath steel surface and they were with sizes over the critical size, the inclusions would greatly influence the fatigue life of steel. Otherwise, the inclusions slightly affect the fatigue behaviors of steel materials. Naturally, larger inclusions would have more important effect on fatigue strength of steel [25, 26]. Also, harder inclusions indicated poorer deformability in hot rolling would have more serious influence on the fatigue resistance property of steel. Usually, critical sizes of inclusions would depend on the locations of them in steel. Commonly, inclusions near the surface or subsurface of steel parts are more detrimental [27–29].

Besides, morphologies of inclusions have important effects on the stress concentrations initiated by them inside steel and therefore, significantly affect fatigue resistance property of steel. Usually, with the rise of inclusions radius, because of smaller curvature, stress concentration initiated by inclusions would be relieved. As a result, negative effects of inclusions can be decreased. Moreover, inclusions in angular or blocky shapes would be more detrimental than inclusions in spherical shape because of easier initiation of stress concentrations inside steel matrix [30]. Also, it is found that mechanical property of steel along transverse and longitudinal directions would be more uniform when the morphology of inclusions was changed from strips into spherical balls.

Based on above description and explanations on the relationship of inclusions natures and anti-fatigue property of steel, it can be summarized that smaller inclusions with spherical shape and low melting points are desirable for enhanced fatigue resistance property of steel.

1.3 Influence of Refining Slag on Steel Cleanliness and Control of Inclusions

As it known, refining slag is composed of multiple components and would often evolve into more complex system after the melting, which in turn would influence various properties of slag. Viscosity is an important proptty of refining slag that greatly affect slag-steel chemical reactions and thus the cleanliness of steel. Too high viscosity of slag is unfavorable to the chemical reactions of slag-steel. Whereas, too lower viscosity would often result in more serious erosion of the linings of metallurgical vessels. In the industrial steelmaking practice, stable viscosity of slag is very significant, viz., viscosity of slag should not be sensitive to the changes of temperature. Hence, in the phase diagram of slag system, the region featured by dense iso-viscosity lines is not the proper choice. Surface tension is another important property of slag that importantly influence the kinetics of slag-steel reactions and the removal of inclusions. Usually, surface tension of refining slag relates to its chemical compositions as well as the atmosphere.

Basicity of slag is commonly referred to the weight percentage ratio of CaO to SiO_2, which affects the ability of slag in removing impurities in steel melts such as phosphorus and sulfur. Higher basicity of slag often favors the dephosphorization and desulfurization of liquid steel thus important to enhance steel cleanliness. Reduction ability of slag is also very important for higher cleanness of steel, which can be evaluated by the contents of unstable oxides including FeO, MnO and Cr_2O_3 in the slag. Usually, basic slag with stronger reduction ability (for example, with total amount of unstable oxides controlled within 1%) should be used for decreased content of oxygen content in steel [31].

1.3.1 Chemical Reactions Among Slag-Steel-Inclusions

During the laboratorial study on control of inclusions, slag-steel reaction experiments are often carried out. Usually, it would be difficult to achieve complete chemical reaction equilibrium among slag, steel and inclusion. However, local equilibrium among them can be possibly established. As a result, slag-steel reactions can be a helpful way to evaluate the control of inclusions in steel and provide important reference for the industrial production. In the experiments, time needed for slag, steel and inclusions to equilibrate with each other should be determined. For example, during the chemical reaction experiment between $CaO–SiO_2–Al_2O_3$ slag and steel and in Ca–O deoxidation, Suito and Tamura determined the reaction time by evaluating the variations of [Al], [Si], [O] and [N] in steel melt, as shown in Fig. 1.4 [32, 33]. It was found that oxygen conent became constant after 60 min' reaction and concentrations of other elements also slightly changed after that. As a resut, equilbrium time for slag and steel reaction were consided as 60 min.

Lee and Suito studied the activity of Fe_tO in $CaO–SiO_2–Al_2O_3–Fe_tO$ slag [34]. They found that when local equilibrium is established for the chemical reaction of slag-steel, chemical equilibrium relationship between Fe_tO in slag and deoxidizing element [Si] in steel would be established and can be deduced according to the chemical reaction equation.

$$3(\mathrm{Fe_tO}) + 2[\mathrm{Al}] = (\mathrm{Al_2O_3}) + 3[\mathrm{Fe}] \tag{1.2}$$

Based on the equilibrium constant of chemical reaction and principles in physical chemistry, the following formula can be deduced:

$$\log(\mathrm{Fe_t}O) = (-2/3)\log[\%\mathrm{Al}] - \log\gamma_{\mathrm{Fe_tO}} + \log\{\alpha \cdot a_{\mathrm{Al_2O_3}}^{1/2})/(K_1^{1/3} \cdot f_{\mathrm{Al}}^{2/3})\} \tag{1.3}$$

In this Eq. (1.3),

α is a constant related to slag composition;
f is the Henry activity coefficient;
γ is the Renault activity coefficient;
K is the equilibrium constant of chemical reaction.

According to Eq. (1.3), it can be known that when the slag compositions are determined, the logarithm of Fe_tO and [Al] contents should be in good linear relationship

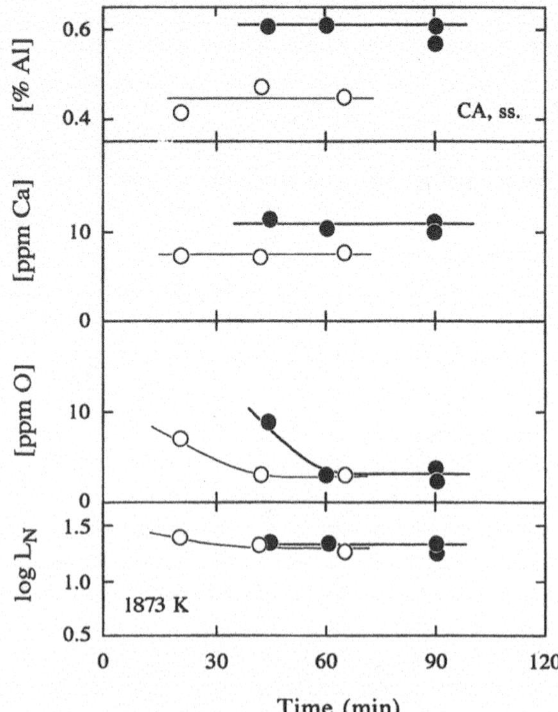

Fig. 1.4 Determination of equilibrium time for slag-steel reaction [33], reprinted by permission from springer nature, Metallurgical and Materials Transactions B, Calcium deoxidation equilibrium in liquid iron, Takashi Kimura and Hideaki Suito, copyright 1994

and slope of the line should be – 2/3. The obtained results showed that experimental data agreed well with theoretical predictions. After the reaction, the logarithm of [Al] content in molten steel and the logarithm of Fe_tO content in slag followed a good linear relationship with slope about −2/3, as shown in Fig. 1.5.

Similarly, for the slag-steel-inclusion system approached chemical reaction equilibrium, as long as the chemical reaction between a specific solute in liquid steel and a component in slag, such linear relationship should exist and were reproducibly proved by Suito H for many times in their studies [35]. When chemical reaction between molten steel and slag approached equilibrium, following relationship among [Al], [Si] in molten steel and (SiO_2), (Al_2O_3) in slag can be deduced.

$$4[\mathrm{Al}] + 3(\mathrm{SiO_2}) = 3[\mathrm{Si}] + 2(\mathrm{Al_2O_3}) \tag{1.4}$$

$$K = \frac{a_{[\mathrm{Si}]}^3 \cdot a_{\mathrm{Al_2O_3}}^2}{a_{[\mathrm{Al}]}^4 \cdot a_{\mathrm{SiO_2}}^3} \tag{1.5}$$

Taking logarithms of both sides at the same time, following equation can be obtained.

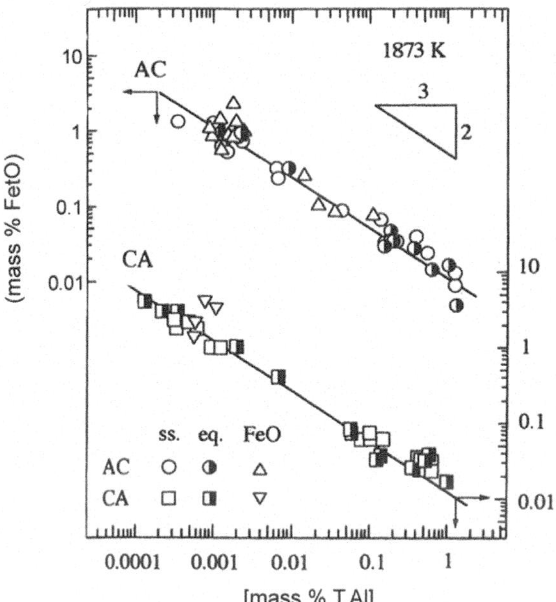

Fig. 1.5 Relation of [T. Al] and (Fe_tO), reprinted by permission from springer nature, Metallurgical and Materials Transactions B, Activities of Fe_tO in $CaO–Al_2O_3–SiO_2–Fe_tO$ (5 pct) slags saturated with liquid iron, Kwang Lo Lee and Hideaki Suito, copyright 1994 [34]

$$\log[\text{Al}] = \frac{3}{4}\log[\text{Si}] + \frac{1}{2}\log a_{\text{Al}_2\text{O}_3} - \frac{3}{4}\log a_{\text{SiO}_2} - \frac{1}{4}\log K \tag{1.6}$$

Since slag composition and reaction temperature are determined and K is a constant at certain temperature, relationship between the logarithm of [Al] and [Si] contents in steel should be a straight line with a slope of 3/4 when this chemical reaction approached thermodynamic equilibrium. Therefore, whether the reaction reaches equilibrium or not can be determined and verified by the relationship between logarithm of actual [Al] and [Si] contents in liquid steel, as shown in the following Fig. 1.6.

Park et al. analyzed the relationship among the chemical compositions of inclusions, liquid steel and slag in ferritic stainless steel by laboratorial experiments, during which the $CaO–MgO–Al_2O_3$ slag was used [36]. The results showed that mole fraction ratio of MgO to Al_2O_3 in the formed magnesia alumina spinel inclusions indicated a close relationship with the compositions of molten steel and slag, as shown in Figs. 1.7 and 1.8.

1.3.2 Influence of Slag Compositions on Steel Cleanness

As mentioned above, complete chemical reaction equilibrium among slag, steel and inclusions was difficult to establish. However, in the actual metallurgical process, local equilibrium of them can be possible. Hence, by proper choices of deoxidation

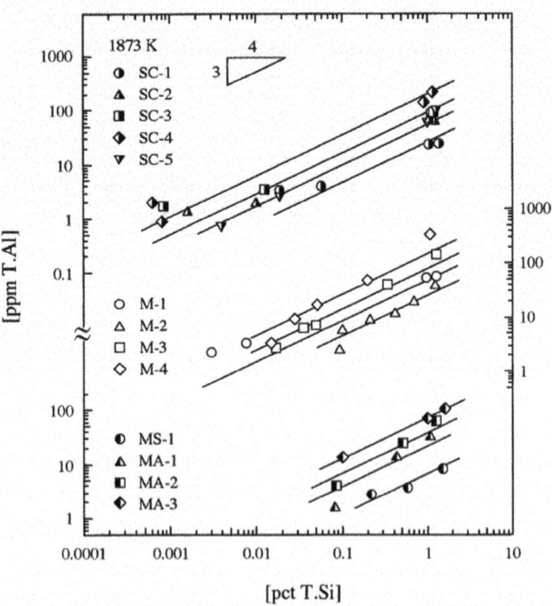

Fig. 1.6 Relation of [T. Si] and [T. Al], reprinted by permission from springer nature, Metallurgical and Materials Transactions B, Activities of SiO_2 and Al_2O_3 and activity coefficients of Fe_tO and MnO in $CaO–SiO_2–Al_2O_3–MgO$ slags, Hiroki Ohta and Hideaki Suito, copyright 1998 [35]

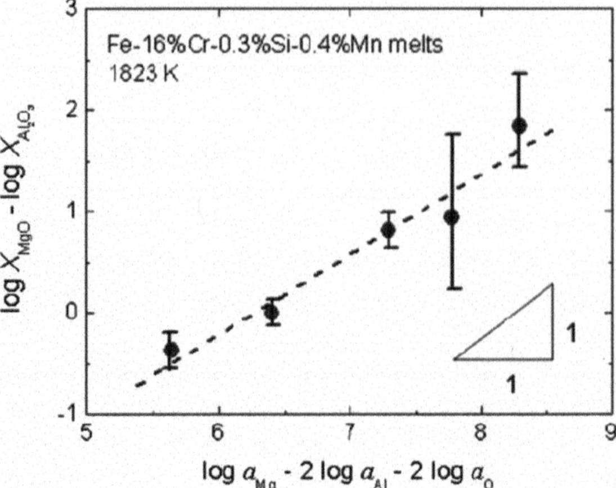

Fig. 1.7 Influence of steel chemistry on inclusion, reprinted by permission from springer nature, Metallurgical and Materials Transactions B, Effect of $CaO–Al_2O_3–MgO$ slags on the formation of $MgO–Al_2O_3$ inclusions in ferritic stainless steel, Joohyun Park and Dongsik Kim, copyright 2005 [36]

and refining slag, chemical compositions of inclusions can be well targeted and controlled.

It has been found that the purity of molten steel closely related to the compositions of refining slag. For example, reducing the content of Fe_tO in slag would be beneficial to reduce the total oxygen content in steel. In order to reduce the content of T[O] in steel to about 0.0010%, the content of Fe_tO in slag should be lower than 0.5%. When the content of Fe_tO in slag is more than 0.5%, the content of T[O], viz., amounts of inclusions in steel would significantly increase [37]. Zhang provided the relationship

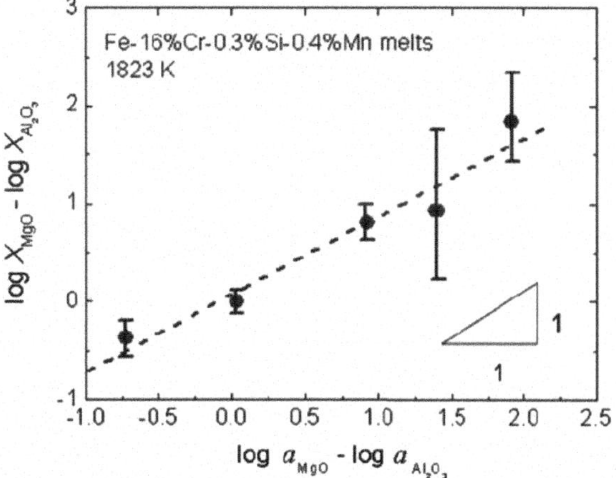

Fig. 1.8 Influence of slag composition on inclusion, reprinted by permission from springer nature, Metallurgical and Materials Transactions B, Effect of CaO–Al_2O_3–MgO slags on the formation of MgO–Al_2O_3 inclusions in ferritic stainless steel, Joohyun Park and Dongsik Kim, copyright 2005 [36]

between the content of Fe_tO in slag and contents of T[O] in steel under different ladle lining and stirring conditions, as shown in Figs. 1.9 and 1.10 [38]. As can be seen, with the increase of Fe_tO content in slag, T[O] contents in steel increased. With the increase of Fe_tO contents in ladle slag, the contents of dissolved aluminum in steel decreased. Also, with the increase of Fe_tO contents in slag, contents of Al_2O_3 in inclusions increased linearly.

Yan et al. reported that sulfur partition ratio (Ls) between slag and steel was strongly affected by the contents of (FeO + MnO) in slag. If the contents of (FeO + MnO) were less than 1%, Ls would significantly increase. By comparison, when the contents of (FeO + MnO) in slag was decreased from 1 to 0.2%, L_s increased by

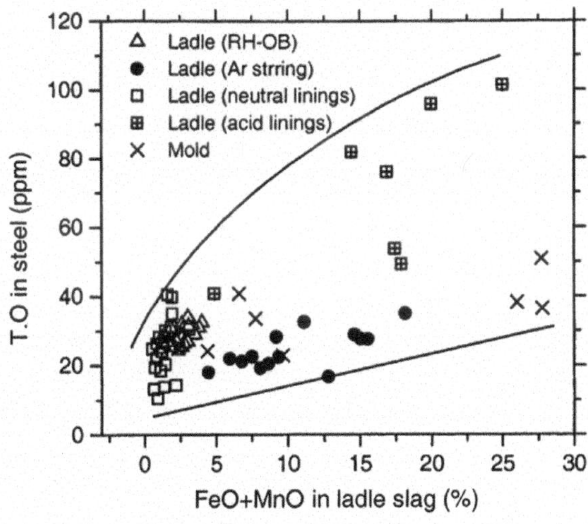

Fig. 1.9 Effect of (FeO + MnO) in slag on T[O] in steel, reprinted from Ref. [38], copyright 2003, with permission from ISIJ

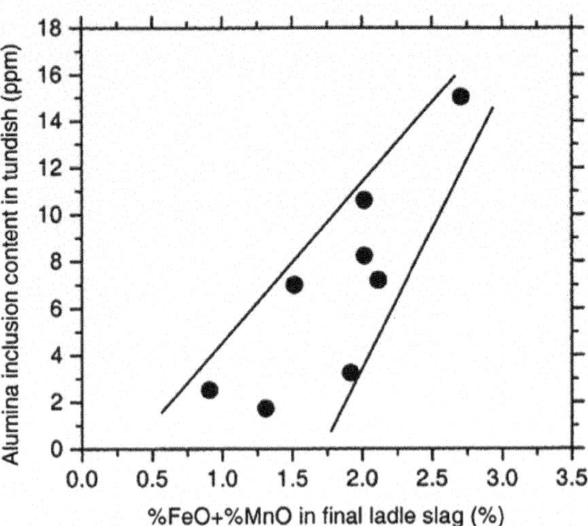

Fig. 1.10 Effect of (FeO + MnO) on alumina content in inclusions, reprinted from Ref. [38], copyright 2003, with permission from ISIJ

5–10 times. When the contents of (FeO + MnO) were higher than 2%, Ls would be very small [39, 40].

Previous studies showed that increasing the basicity of slag would be favorable for deoxidation and desulfurization of molten steel [41, 42]. Relevant research of bearing steel showed that total oxygen contents in steel would be decreased when the slag basicity was increased from 3.5–4.5 to 4.5–7.0 [43]. Moreover, the weight percentage ratio of CaO/Al_2O_3 in slag also indicated great influence on T[O] contents in steel. When the ratios of CaO/Al_2O_3 in slag were decreased from 2.0–4.4 to 1.2–2.0, T[O] contents were reduced from about 12×10^{-6} to 8×10^{-6}, as shown in Fig. 1.11. The results revealed that when the contents of CaO in slag was constant, increasing Al_2O_3 content in slag was beneficial to deoxidation.

Sulfur capacity of slag is often evaluated in the investigations of desulfurization. It had been found that changes in slag compositions would lead to the variations

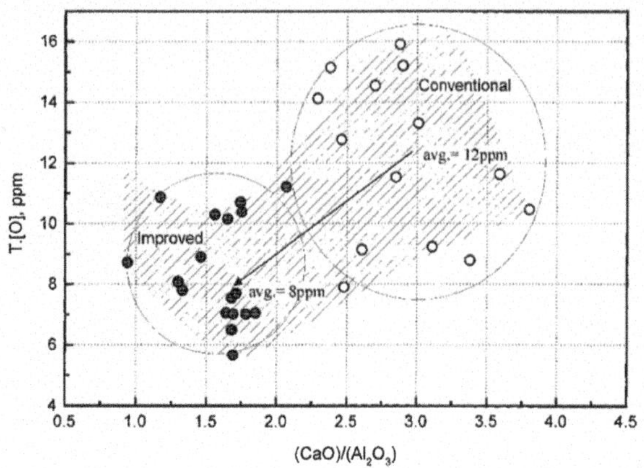

Fig. 1.11 Effect of CaO/Al_2O_3 ratios in refining slag on T[O] contents in steel, reprinted from Ref. [43], copyright 2002, with permission from Taylor & Francis

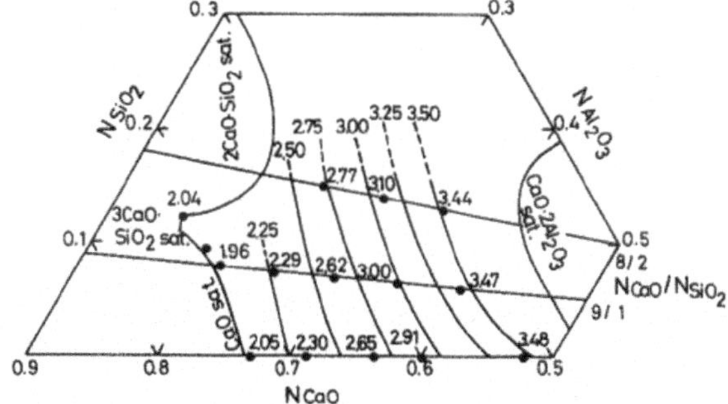

Fig. 1.12 Iso-logC_s lines in CaO–Al_2O_3–SiO_2 slag, reprinted from Ref. [44], copyright 1993, with permission from ISIJ

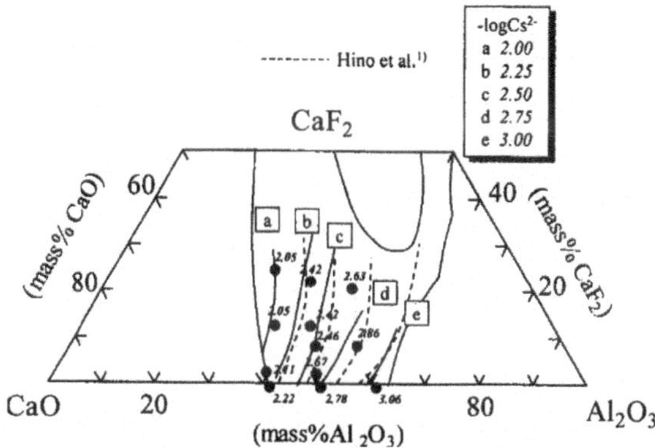

Fig. 1.13 Iso-logC_s lines in CaO–CaF_2–Al_2O_3 slag, reprinted from Ref. [45], copyright 2003, with permission from ISIJ

in sulfur capacity of slags, which ultimately affected the desulfurization ability of slag. For example, the study of Hino showed that when the contents of Al_2O_3 in slag remained stable, addition of SiO_2 and MgO to CaO–Al_2O_3 slag while decreasing CaO content at the same time would reduce the sulfur capacity of slag. Moreover, sulfur capacity of slag with the addition of SiO_2 was smaller than slag with the addition of MgO, as can be seen in Figs. 1.12 and 1.13 [44]. Ohta et al. studied the effect of CaF_2, MgO and SiO_2 on the sulfur capacity of CaO–Al_2O_3 slag [45]. The results showed that when CaO content in the slag was very high, addition of MgO would significantly improve sulfur capacity of the CaO–Al_2O_3 slag, while the addition of SiO_2 would reduce the sulfur capacity of the CaO–Al_2O_3 slag, as shown in Figs. 1.14 and 1.15.

Proper choice of refining slag is also meaningful in achieving lower oxygen content in steel. For example, in the industrial production practice of bearing steel, Sanyo Special Steel Co. Ltd. adopted high basicity refining slag, as given in Table 1.1 [46]. Mass fraction of sulfur content in steel can be reduced to 20×10^{-6}–30×10^{-6} while the oxygen content can be reduced to 3×10^{-6}–4×10^{-6}. After the refining, the ASTM A-type and B-type inclusions can hardly be found in steel while the formed inclusions were mainly the ASTM D-type.

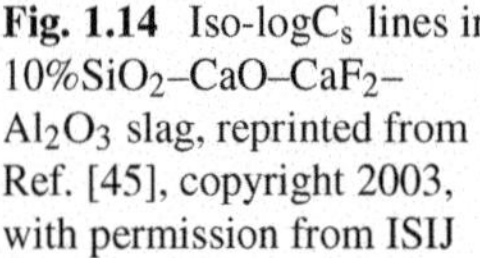

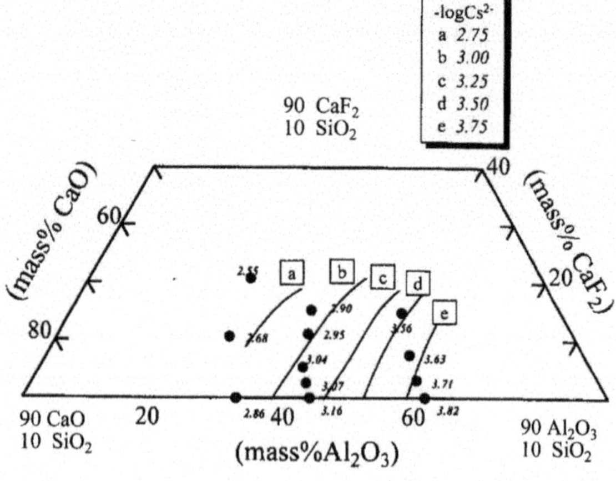

Fig. 1.14 Iso-logC_S lines in 10%SiO_2–CaO–CaF_2–Al_2O_3 slag, reprinted from Ref. [45], copyright 2003, with permission from ISIJ

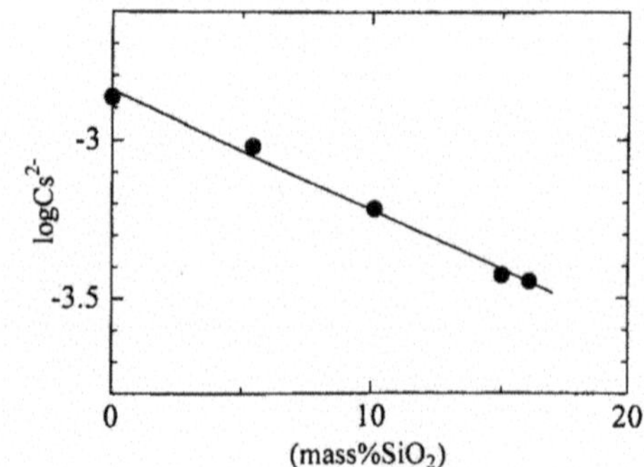

Fig. 1.15 Effect of SiO_2 on sulfur capacity, reprinted from Ref. [45], copyright 2003, with permission from ISIJ

Table 1.1 Composition of refining slag of Sanyo special steel, mass%

Slag compositions, mass%									
CaO	SiO_2	Al_2O_3	MgO	MnO	TFe	Cr_2O_3	P_2O_5	CaF_2	S
57.8	13.3	15.8	4.3	< 0.1	0.6	< 0.1	< 0.1	7.8	1.1

During the secondary refining, changes of Al_2O_3 contents in slag showed great influences on the cleanness of molten steel. On one hand, variations of Al_2O_3 contents in slag would directly affect the activities of Al_2O_3 in slag thus the chemical reaction equilibrium of Al–O deoxidation. On the other hand, changes in Al_2O_3 contents in slag would influence the physical and chemical properties of slag, thus affecting the absorption of inclusions by slag.

The Al_2O_3 contents also largely influenced the desulfurization ability of refining slag. On one hand, with the increase of Al_2O_3 contents, viscosity of slag would decrease. Hence, kinetic conditions of desulfurization would be influenced. On the other hand, the increase of Al_2O_3 contents in slag would reduce the activity of CaO in

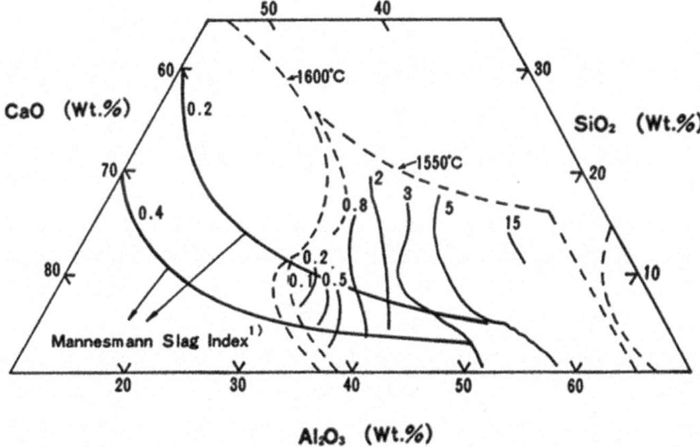

Fig. 1.16 The S.P. index lines of $CaO–Al_2O_3–SiO_2$ slag, reprinted from Ref. [50], copyright 1986, with permission from ISIJ

slag and thus inhibit desulfurization of liquid steel [47–49]. In recent years, refining slag composed of $CaO–Al_2O_3$ is widely used in secondary refining.

Ogura et al. proposed the S.P. index to evaluate the desulfurization ability of refining slag, which can be expressed by the following Eq. (1.7) [50]. And they measured the variations of S.P. index for $CaO–SiO_2–Al_2O_3$ system slag in different chemical compositions. The results showed that the S.P. index of slag would be the lowest when CaO content is 60%, SiO_2 content is 10%, and Al_2O_3 content is 30%, as shown in Fig. 1.16.

$$(\text{S.P.}) = \frac{a_{\text{Al}_2\text{O}_3}^{1/3}}{a_{\text{CaO}} \cdot (\%\text{S})_{\text{sat}}} \tag{1.7}$$

$$\frac{(\%\text{S})}{[\%\text{S}]} = \frac{a_{\text{CaO}} \cdot (\%\text{S})_{\text{sat}}}{a_{\text{Al}_2\text{O}_3}^{1/3}} \cdot [\%\text{Al}]^{2/3} \cdot K_{\text{s}} \tag{1.8}$$

1.3.3 Influence of Slag Compositions on the Control of Inclusions

Slag compositions also indicated important influences on the control of non-metallic inclusions in steel. On one hand, variations in slag compositions influenced the absorption of inclusions by slag. As it known, inclusions can be floated to the slag-steel interface during the refining, removed from steel by dissolving into the slag. On the other hand, slag importantly affects the compositions of inclusions by chemical reactions among slag-steel-inclusions.

Removal of inclusions from steel-slag interface into slag phase closely related to the interface interaction among slag-steel-inclusions. It had been known that several steps would be involved during the absorption of inclusions by slag. Firstly, the inclusions were separated from steel to the interface of steel-slag. Secondary, the inclusions would be transferred from the interface into slag. Thirdly, the inclusions can dissolve well and enter into the slag.

Emi and Yin et al. had conducted classical in-situ observations at high temperature on the behaviors of inclusions in liquid steel and at the steel-slag interface by the high temperature laser scanning microscopy [51–60]. The results showed that behaviors of inclusions at steel-slag interface were obviously affected by the surface interaction between inclusions and steel as well as slag. For example, Al_2O_3 inclusions showed strong tendency to agglomerate into clusters because of poor wettability in liquid steel. However, other liquid inclusions indicated weak agglomeration tendency.

As a result, if inclusions can be removed by the absorption of slag, it should firstly move to the steel-slag interface [61]. Then, whether the inclusions can enter into the slag phase or remain in steel depends on the interactions between the inclusion and the interface. Cramb and Jimbo pointed out that after inclusions enter the interface, they did not immediately enter into slag phase but stay at the interface [62]. Up to now, many works have been done to reveal the removal of inclusions based on thermodynamics [63–66].

A series of works have been done to study the dissolution behaviors of Al_2O_3, MgO and $MgO{\cdot}Al_2O_3$ inclusions into different slags. For example, Nakajima and Okamura established a mathematical model to study the behaviors of inclusions at the interface between steel and slag [67]. The model considered the behaviors of inclusions under two conditions: (1) when there is a layer of liquid steel film between inclusions and slag; (2) when there is no liquid steel film between inclusions and slag. Strandh et al. established a mathematical model and proved that interfacial tension and slag viscosity played very important roles in the separation of inclusions from steel and entering into slag phase, as shown in Figs. 1.17 and 1.18 [68]. In order to ensure the dissolutions of inclusions into slag phase, slag viscosity must be relatively low and inclusions should be well wetted by the slag.

Choi et al. studied the dissolution rate of Al_2O_3 inclusions in $CaO–Al_2O_3–SiO_2$ slag [69]. It was found that dissolution rate of Al_2O_3 inclusions in the ternary system

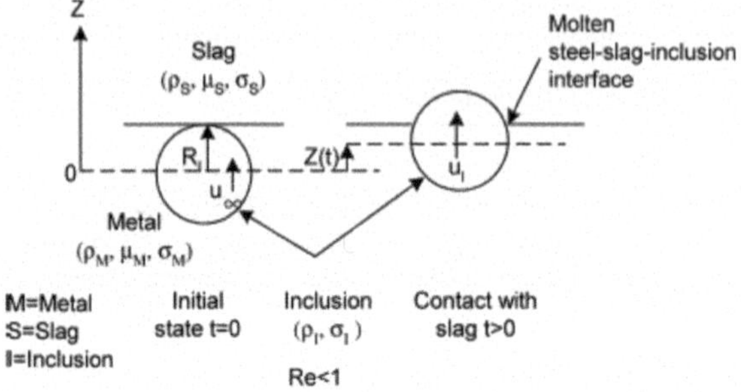

Fig. 1.17 Schematics on the behaviors of inclusions at slag-steel interface (without steel film), reprinted from Ref. [68], copyright 2005, with permission from ISIJ

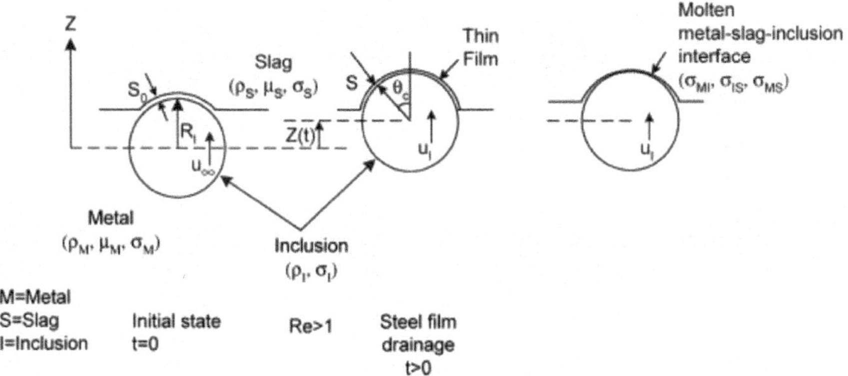

Fig. 1.18 Schematics on the behaviors of inclusions at slag-steel interface (with steel film), reprinted from Ref. [68], copyright 2005, with permission from ISIJ

slag depended on the following factors: (1) with the rise of temperature, dissolution rate of Al_2O_3 inclusions was increased; (2) with the increase of CaO contents in slag, dissolution rate of Al_2O_3 inclusions increased; (3) at a fixed CaO content, the dissolution rate of Al_2O_3 inclusions in slag increased. Moreover, the dissolution rate of inclusions into slag also depended on the viscosity of slag and the diffusion ability of inclusions in slag. They obtained the relative equal dissolution rate diagram of Al_2O_3 in CaO–Al_2O_3–SiO_2 slag (57.7%CaO–42.3%Al_2O_3). As can be seen from Fig. 1.19, contents of CaO in the slag largely affected dissolution rate of Al_2O_3 in CaO–Al_2O_3–SiO_2 slag.

Fig. 1.19 Dissolution rate of Al_2O_3 in the CaO–Al_2O_3–SiO_2 slag, reprinted from Ref. [69], copyright 2002, with permission from ISIJ

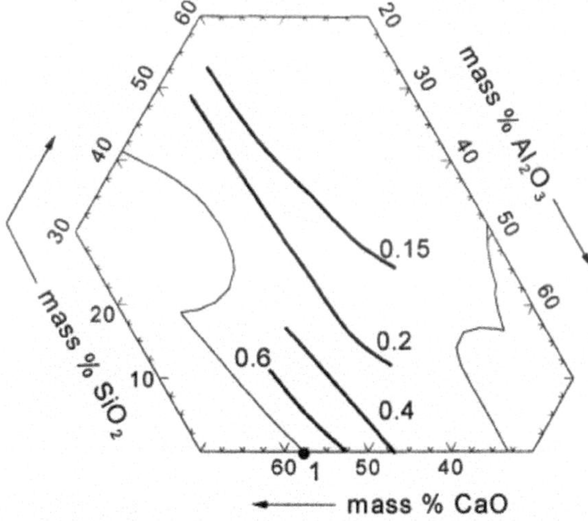

In addition, Choi and Lee also studied the wettability of $CaO–Al_2O_3–SiO_2$ slag on solid Al_2O_3 inclusions [70]. They established mathematical model to reveal the relationship of slag-inclusion contact angle and dissolving time. During the modeling, continuous chemical interaction between inclusions and slag was taken into account. And curve of contact angles between liquid slag and Al_2O_3 inclusions was obtained based on the experimental data and calculation results. The results showed that composition of slag had great influences on the contact angle. With the increase of SiO_2 contents in slag, high basicity slag has good wettability to Al_2O_3. When the basicity of slag was fixed, the rise of Al_2O_3 contents in slag, it would be more difficult for the wetting of inclusions by slag. At a fixed CaO/Al_2O_3 mass ratio, changes of SiO_2 contents slightly effected the wettability of inclusions by slag. This can be read from Figs. 1.20 and 1.21.

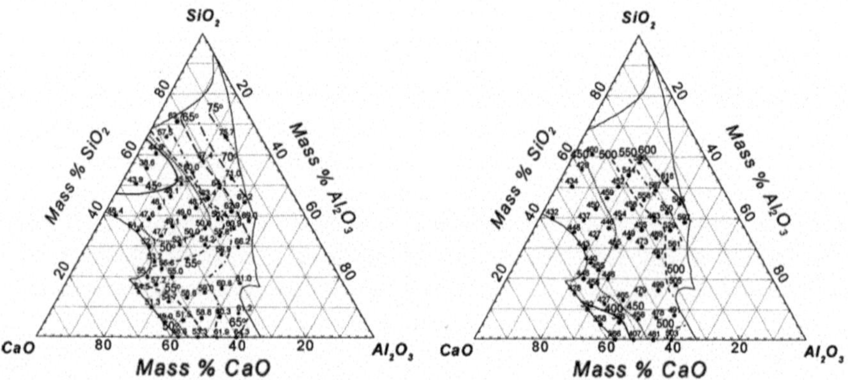

Fig. 1.20 Iso-contact angle and iso-surface tension lines of Al_2O_3 and $CaO–Al_2O_3–SiO_2$ slag, reprinted from Ref. [70], copyright 2003, with permission from ISIJ

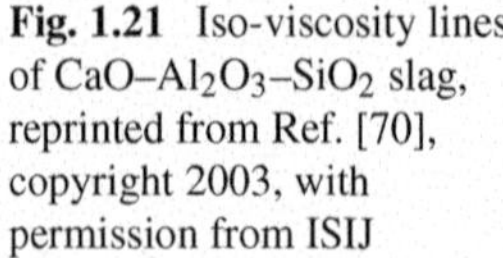

Fig. 1.21 Iso-viscosity lines of $CaO–Al_2O_3–SiO_2$ slag, reprinted from Ref. [70], copyright 2003, with permission from ISIJ

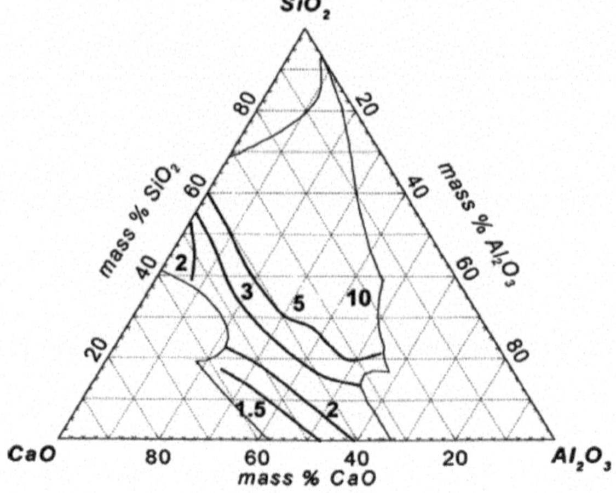

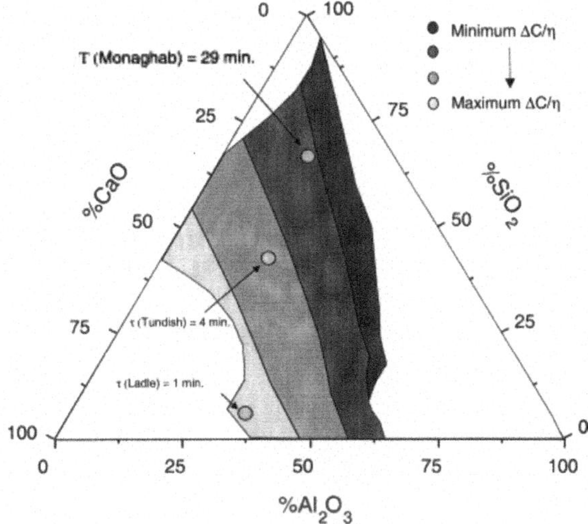

Fig. 1.22 $\Delta C/\eta$ in $CaO–Al_2O_3–SiO_2$ slag, reprinted from Ref. [71], copyright 2006, with permission from ISIJ

Valdez et al. studied the absorption capacity of $CaO–Al_2O_3–SiO_2$ slag for oxide inclusions [71]. It was found that even if there was a liquid steel film between slag and inclusions, it slightly affected the dissolution of inclusions into slag. The results showed that: (1) behaviors of inclusions at the interface of steel-slag-inclusions mainly depended on the interfacial energy, and also affected by the slag viscosity; (2) effect of the film can be ignored generally; (3) dissolution rate and supersaturation of inclusions were proportional to slag viscosity. According to the absorption capacity of slag to Al_2O_3 inclusions, liquid phase of $CaO–Al_2O_3–SiO_2$ slag can be divided into four regions, as shown in Fig. 1.22. Viscosities of high-temperature slag systems can also be estimated by the mathematical model proposed by Seetharaman and Du [72], or turn to the Handbook edited by Kawayi and Shiraishi [73].

Besides, refining slag also indicated important influences on chemical compositions of inclusions. During the establishment of chemical reaction equilibrium between the slag and strong deoxidizing elements in steel such as [Ca], [Mg], [Al] etc., composition of inclusions would be greatly influenced. The relationship among liquid steel, slag and inclusions can be summarized as follows: slag composition directly influenced the compositions of liquid steel, which in turn affected the compositions of inclusions. As a result, the needed chemical compositions of liquid steel and refining slag to target proper inclusions can be predicted based on thermodynamic calculations.

In order to calculate the formation thermodynamics of inclusions, it is necessary to measure or get the activity data of slag components under varied slag compositions. Suito et al. had done a series of fundamental works in this field and the involved slag systems included $CaO–SiO_2$, $CaO–Al_2O_3$, $CaO–MgO–Al_2O_3$, $CaO–SiO_2–Al_2O_3$ and $MnO–Al_2O_3–SiO_2$ and so on [35, 74–78].

For example, Suito et al. and Hagiwara et al. conducted thermodynamic calculation on the control of non-metallic inclusions in valve spring steel, bearing steel,

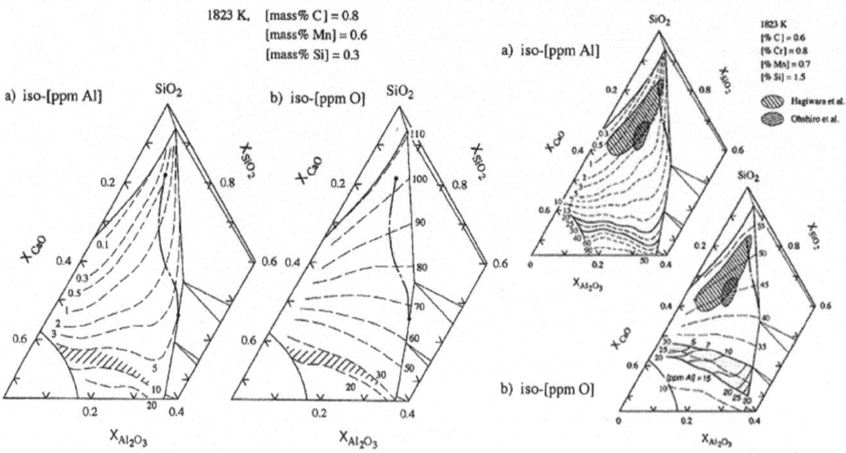

Fig. 1.23 Iso-[Al] and iso-[O] lines of spring steel equilibrated with CaO–Al_2O_3–CaO slag at 1823 K, reprinted from Ref. [79], copyright 1996, with permission from ISIJ

cord steel and other steel grades of automobile engine, [79, 80] as shown in Fig. 1.23. In order to target inclusion with plastic deformability in valve spring steel, slag with low basicity and low alumina content must be used. And the [Al] contents in liquid steel must be strictly controlled within very low level of about 0.5×10^{-6}–5×10^{-6} [80].

1.4 Control of Calcium Magnesia Aluminate Inclusions in Steel

In the process of steelmaking, molten steel contacted with slag and refractories of metallurgical vessels. Usually, slags are composed CaO, MgO, Al_2O_3, SiO_2 and other components while the MgO-based refractory is the popular choice. For Al killed steel, the formed inclusions would be usually composed of calcium magnesia aluminate. As mentioned above, deformability of inclusions closely related to their melting temperatures. So, to relieve negative effects of inclusions in Al deoxidized steels, the goal of inclusions is to target low melting temperature calcium magnesia aluminates.

During steelmaking process, calcium treatment and slag refining are the two ways to control the compositions of inclusions and realize low melting point of inclusions in Al killed steels.

1.4.1 *Influence of Calcium Treatment on the Control of Inclusions*

As it known, alumina inclusions would be largely produced in steel after Al deoxidation, which were featured by high melting points. During continuous casting, Al_2O_3 inclusions like to deposit on the inner wall of submerged nozzles to cause clogging problems. In order to relieve the problem, calcium treatment was introduced to transform Al_2O_3 into low melting point calcium aluminate [81–85].

To elucidate the modification mechanism of alumina into calcium aluminates, many studied have been carried out. Ye et al. proposed the un-reaction core model to explain the transformation of Al_2O_3 inclusions during calcium treatment [86]. In steelmaking practice, the intensity of calcium treatment is often evaluated by the [Ca]/[S] ratios. In Fig. 1.24, changes of inclusions in calcium treatment were schematically described [87].

If calcium treatment was properly conducted, chemical compositions of the formed calcium magnesia aluminate inclusions in steel can be controlled in the lower melting point region. In industrial practice, there are some important factors that greatly affect the results of calcium treatment, such as:

(1) During calcium treatment, due to high vapor pressure of calcium at high temperature, the yield of calcium is relatively low and not stable, especially when the slag contained higher (FeO) contents, as shown in Fig. 1.25 [88].

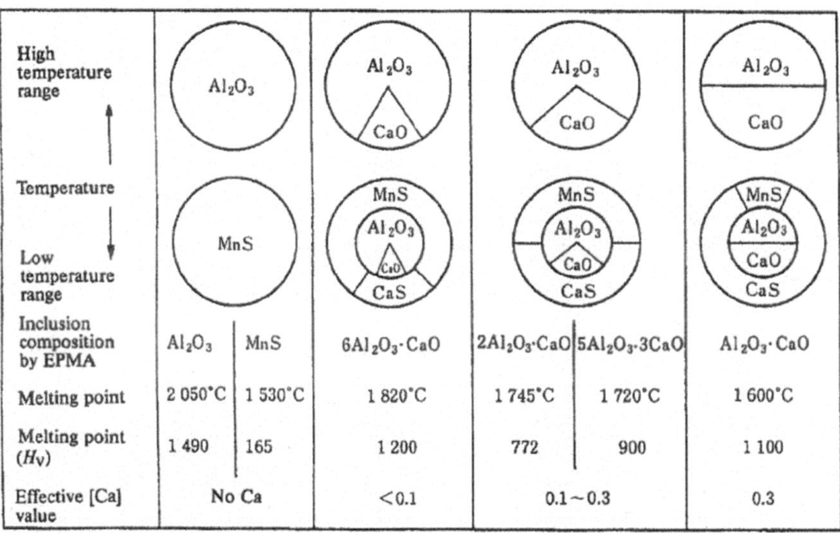

Fig. 1.24 Effect of [Ca]/[S] on the modification of inclusions in Ca-treatment, reprinted from Ref. [87], copyright 1977, with permission from ISIJ

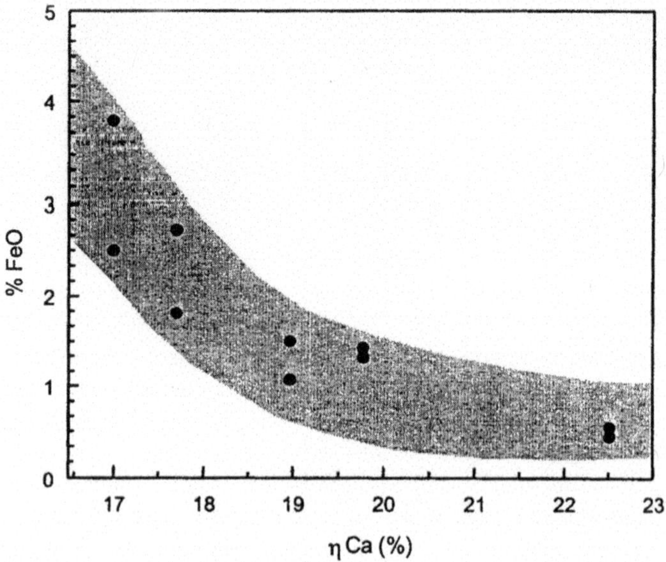

Fig. 1.25 Effect of FeO on the yield ratio of Ca, reprinted from Ref. [88], copyright 2001, with permission from AIST

(2) During calcium treatment, it is not easy to ensure constantly high contents of [Ca] in molten steel. As shown in Fig. 1.26, from the end of the calcium treatment to the continuous casting tundish, although the [Ca] content was about

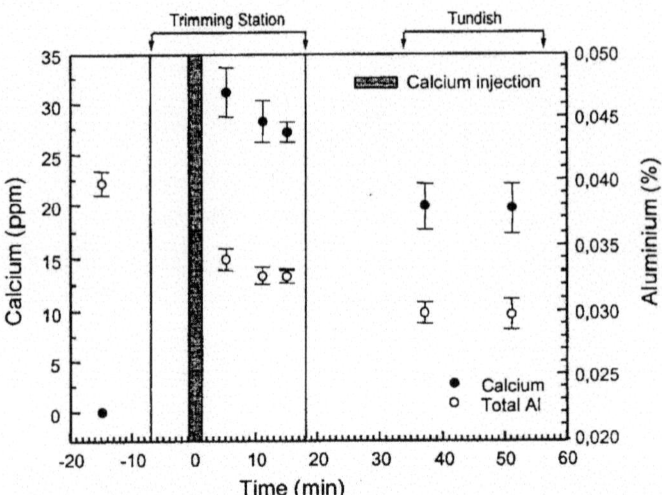

Fig. 1.26 Changes of [Ca] and [Al] in process, reprinted from Ref. [88], copyright 2001, with permission from AIST

0.0030% just after calcium treatment, it decreased obviously and continuously to about 0.0020% [88].

(3) Before calcium treatment, the oxygen and sulfur contents in liquid steel must be controlled in an appropriate and narrow range to avoid the formation of high melting point CaS inclusions in liquid steel, which can also easily cause nozzle clogging.

(4) Calcium treatment would increase the production cost. In addition, in the actual production process, stable modification of inclusion is not easy to absolutely ensure in the massive industrial production.

1.4.2 Influence of Slag-Steel Reaction on the Control of Inclusions

Control of magnesia aluminate inclusions with composition approaching spinel had been a very hot topic in steelmaking in recent decades because of their negative influences on the quality of final products. Due to wide use of MgO-based or MgO-contained materials for lining of vessels and for slag making, a certain amount of Mg would certainly enter into molten steel through the interface and chemical reaction at high temperature. Studies proved that that spinel inclusions can be formed in steel even if the content of [Al] and [Mg] in steel were of a few ppm [89]. Therefore, spinel inclusions are one of the typical inclusions in the production of high-quality steels.

Initially, important attentions were paid to spinel inclusions by metallurgical workers during the production of stainless steels. Due to high MgO contents in AOD lining and slag, even if Al is strictly controlled in various materials during deoxidation of stainless steel, spinel inclusions can be easily formed in stainless steel [90]. Because of high melting point and poor deformation ability, spinel seriously deteriorates the surface quality of stainless steel. For this reason, systematical studies were carried out to spinel inclusions, including the formation thermodynamics calculation and kinetics especially for stainless steel.

Itoh et al. studied the formation of $MgO{\cdot}Al_2O_3$ in steel with complex deoxidation of Al and Mg [91]. The results showed that MgO-based refractory would react with molten steel, resulting in the reduction of MgO in refractories to supply [Mg] into molten steel. Then, [Mg] and [O] in molten steel would react with Al_2O_3 formed in Al deoxidation to produce $MgO{\cdot}Al_2O_3$, which can be expressed as following reaction 1.9 and reaction 1.10.

$$3(MgO) + 2[Al] = 3\left[Mg\right] + (Al_2O_3) \tag{1.9}$$

$$\left[Mg\right] + (Al_2O_3) + [O] = (MgO \cdot Al_2O_3) \tag{1.10}$$

Phase stability diagram of $MgO/MgO{\cdot}Al_2O_3/Al_2O_3$ can be obtained by thermodynamic calculation, as shown in Fig. 1.27. It was also found that if the liquid steel

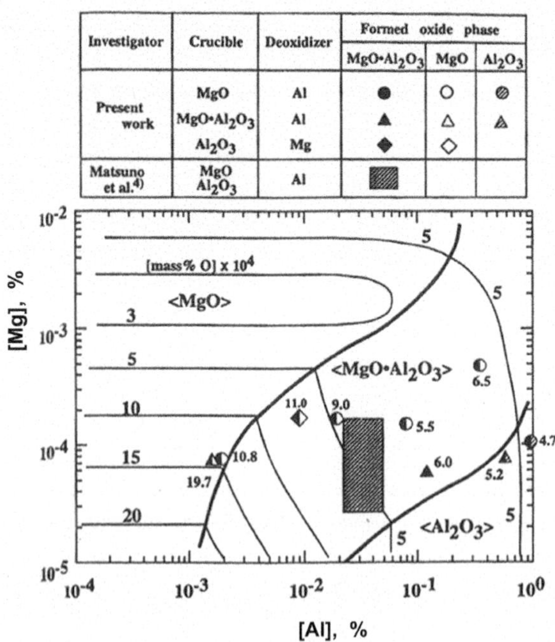

Fig. 1.27 Stability diagram of $MgO/MgO{\cdot}Al_2O_3/Al_2O_3$, reprinted by permission from springer nature, Metallurgical and Materials Transactions B, Effect of $CaO–Al_2O_3–MgO$ slags on the formation of $MgO–Al_2O_3$ inclusions in ferritic stainless steel, Hiroyasu Itoh, Mitsuraka Hino and Shiro Ban-Ya, copyright 1997 [91]

contained certain amount of [Ca], $MgO{\cdot}Al_2O_3$ inclusions would be unstable and modified into calcium aluminates. For example, when the content of [Ca] in steel was as low as 0.0002%, $MgO{\cdot}Al_2O_3$ can be transformed into $CaO{\cdot}2Al_2O_3$, as shown in Fig. 1.28. Kim et al. and Cha et al. investigated the precipitation behavior of spinel in entrapped slag particles of AOD while Nishi et al. studied the effect of refining slag on the formation of spinel inclusions in stainless steel, as shown in Fig. 1.29 [92–95]. They found that spinel would precipitate/crystalize on the entrapped AOD slag particles during the solidification of steel because the entrapped AOD slag particles were often with chemical compositions in spinel-saturated region. To prevent or prohibit the formation of $MgO{\cdot}Al_2O_3$, it was necessary to reduce the amount of Al used in deoxidation or the content of MgO in slag.

Shi et al. studied the formation of spinel in the Al deoxidized 304 stainless steel under the refining of the $CaO–Al_2O_3–SiO_2–MgO$ slag [95]. It was found that content of MgO in spinel inclusions increased with the rise of refining time. With the rise of slag basicity, contents of MgO in inclusions increased. Moreover, with the rise of MgO contents in slag, MgO contents in inclusions would also increase. The maximum amount of MgO contents in spinel inclusions depended on the mass ratios of CaO/SiO_2 in slag, as shown in Figs. 1.30 and 1.31.

Cha et al. studied the formation of $MgO{\cdot}Al_2O_3$ inclusions in ferritic stainless steel contained different contents of [Al] [94]. The results showed that with the transferring liquid steel from AOD to casting tundish, the entrapped slag particles during AOD tapping would partly collide with Al_2O_3 inclusions formed in Al deoxidation, resulting in the increase of Al_2O_3 contents in the entrapped particles. As a result,

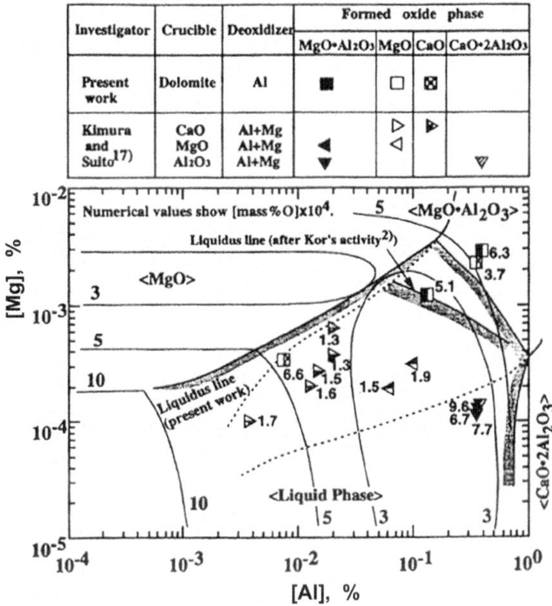

Investigator	Crucible	Deoxidizer	Formed oxide phase			
			$MgO \cdot Al_2O_3$	MgO	CaO	$CaO \cdot 2Al_2O_3$
Present work	Dolomite	Al	■	□	⊠	
Kimura and Suito[17]	CaO MgO Al_2O_3	Al+Mg Al+Mg Al+Mg	◀ ▼	▷ ◁	▶	▽

Fig. 1.28 Stability diagram with [Ca] about 0.0002%, reprinted by permission from springer nature, Metallurgical and Materials Transactions B, Effect of CaO–Al_2O_3–MgO slags on the formation of MgO–Al_2O_3 inclusions in ferritic stainless steel, Hiroyasu Itoh, Mitsuraka Hino and Shiro Ban-Ya, copyright 1997 [91]

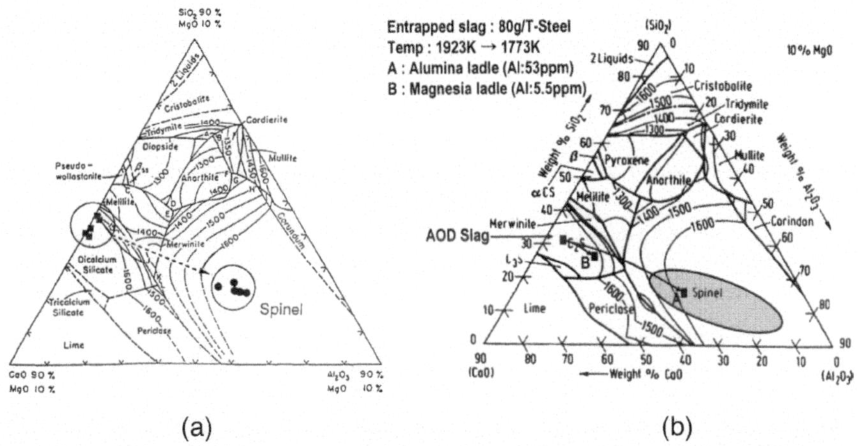

Fig. 1.29 Crystallization of spinel on entrap slag particle in stainless steel: **a** result by Kim et al., reprinted from Ref. [93], copyright 1996, with permission from ISIJ, **b** result by Cha et al., reprinted from Ref. [94], copyright 2004, with permission from ISIJ

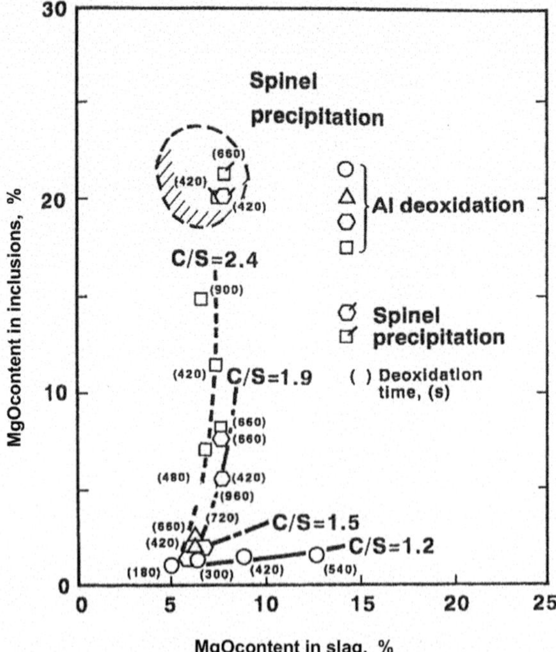

Fig. 1.30 Influence of MgO content in slag on MgO content in inclusions, reprinted from Ref. [95], copyright 1998, with permission from ISIJ

compositions of inclusions can probably enter into the $MgO{\cdot}Al_2O_3$ saturation region. During the solidification of steel, spinel would crystallize on the such entrapped slag particles with the decreasing of temperature. As a result, compositions of such complex inclusions would be close to that of slag. Research of Park also showed that spinel inclusions can precipitate on slag particles involved in tapping process, [96] in which direct evidence on the crystallization of $MgO{\cdot}Al_2O_3$ on entrapped slag particles was provided and proposed the formation mechanism of $MgO{\cdot}Al_2O_3$ in stainless steel, as shown in Figs. 1.32 and 1.33.

Formation of $MgO{\cdot}Al_2O_3$ inclusions in 304 stainless steel melts refined by CaO–MgO–Al_2O_3–SiO_2 slag was also studied by Ehara et al. Their results showed that spinel inclusions can be formed in steel in two ways [97, 98]. The first way was by the direct reaction of [Mg], [Al] and [O] in steel. When the contents of [Mg], [Al] and [O] in steel was in the formation region of $MgO{\cdot}Al_2O_3$, spinel would be produced in steel. They proposed that when the contents of [Al] in steel were lower than 6 ppm, $MgO{\cdot}Al_2O_3$ inclusions can be well prevented in casting billet. While the second way was by the crystallization of spinel on entrapped CaO–MgO–Al_2O_3–SiO_2 slag particles. They also reported that spinel can easily crystallize on such slag particles when their compositions located in the spinel saturation region in the CaO–Al_2O_3–SiO_2–(5%, 10%) MgO system, as shown in Fig. 1.34. Contents of Al_2O_3 in inclusions would increase with the rise of the contents of Al_2O_3 contents in slag. During continuous casting, the rise of Al_2O_3 contents in inclusions often increase due to the re-oxidation of steel, as shown in Figs. 1.35, 1.36 and 1.37. By controlling

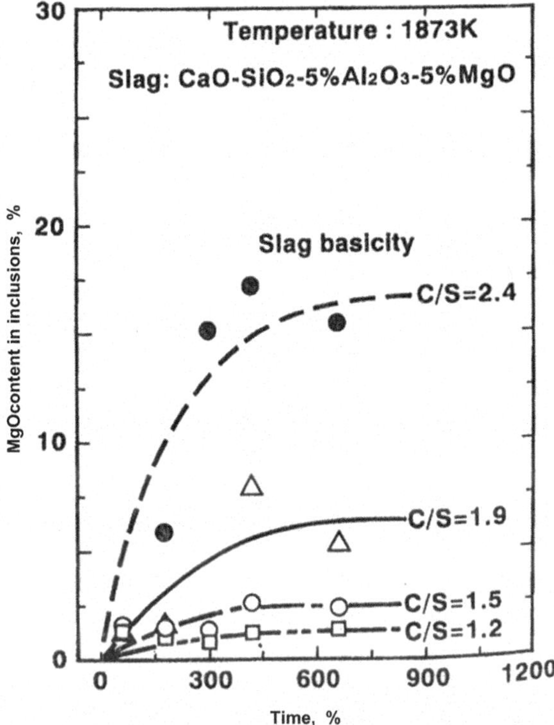

Fig. 1.31 Influence of slag basicity on MgO content in inclusions, reprinted from Ref. [95], copyright 1998, with permission from ISIJ

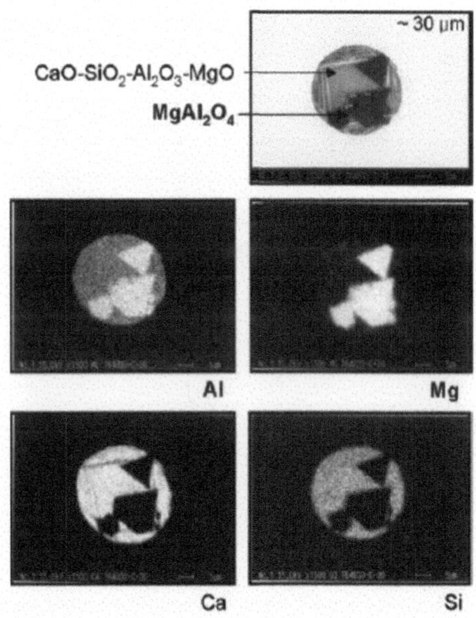

Fig. 1.32 Crystal of $MgO{\cdot}Al_2O_3$ on entrapped slag, reprinted by permission from springer nature, Metallurgical and Materials Transactions B, Formation mechanism of spinel-type inclusions in stainless steel, Joohyun Park, copyright 2007 [96]

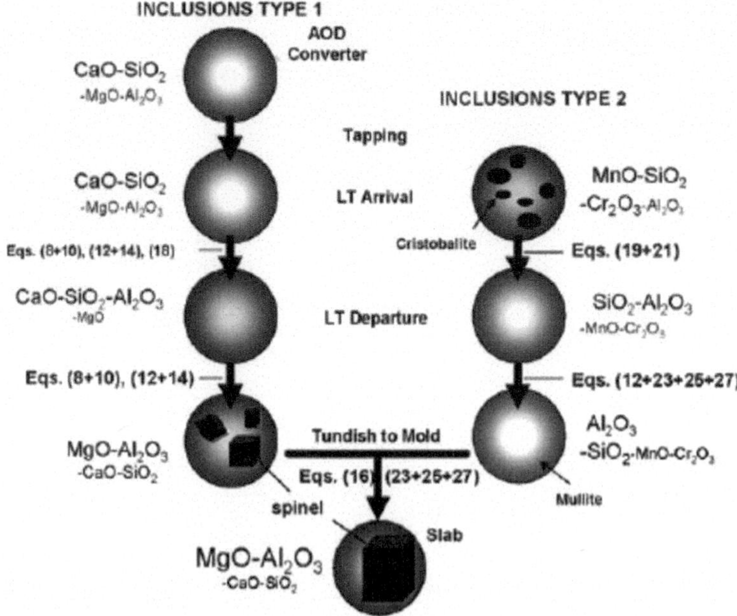

Fig. 1.33 Schematics on the precipitation of spinel, reprinted by permission from springer nature, Metallurgical and Materials Transactions B, Formation mechanism of spinel-type inclusions in stainless steel, Joohyun Park, copyright 2007 [96]

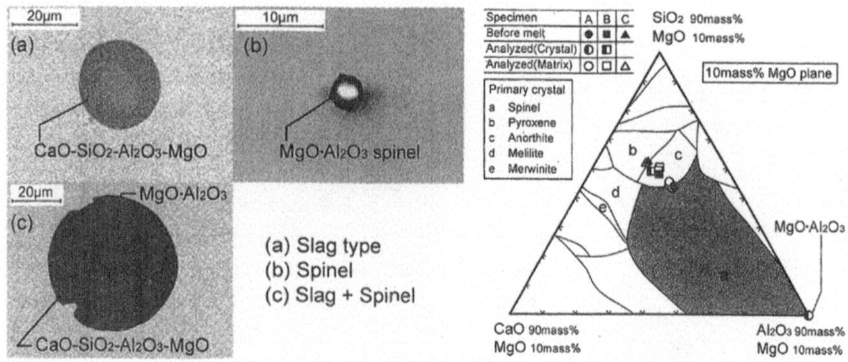

Fig. 1.34 $MgO{\cdot}Al_2O_3$ in stainless steel, reprinted from Ref. [97], copyright 2007, with permission from ISIJ

slag basicity and reducing Al_2O_3 content in slag, formation of $MgO{\cdot}Al_2O_3$ can be prevented, as shown in Fig. 1.38.

Todoroki and Mizuno studied the control of inclusions in Al deoxidized 304 stainless steel melts refined by $CaO–Al_2O_3–MgO–CaF_2$ and $CaO–SiO_2–Al_2O_3–MgO–CaF_2$ slag [99, 100]. It was found that a large number of Al_2O_3 inclusions were

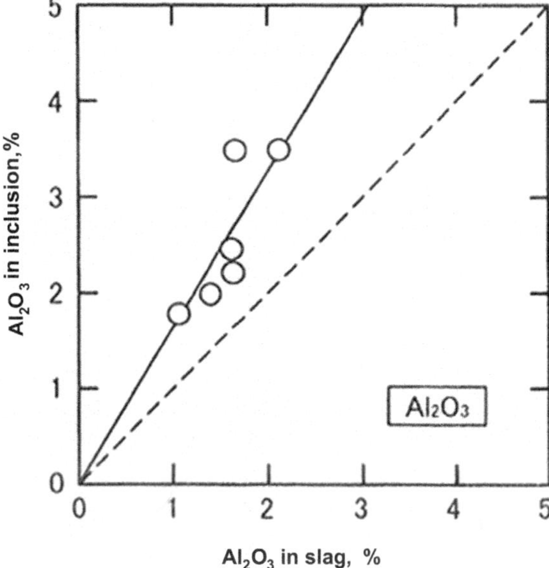

Fig. 1.35 Changes in Al_2O_3 in AOD slag, reprinted from Ref. [98], copyright 2007, with permission from ISIJ

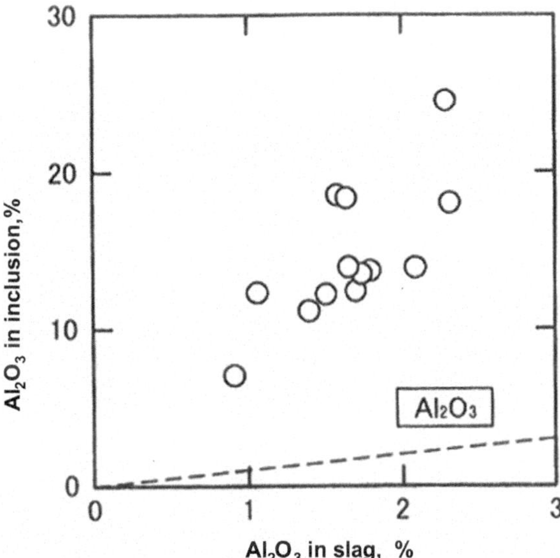

Fig. 1.36 Changes in Al_2O_3 in inclusion in slab, reprinted from Ref. [98], copyright 2007, with permission from ISIJ

formed in steel shortly after the addition of Al. Simultaneously, MgO in slag would be reduced by [Al] in steel to supply [Mg] into steel. Afterwards, chemical reaction between [Mg] and Al_2O_3 inclusions would result in the formation of $MgO{\cdot}Al_2O_3$ inclusions in steel. When the slag was free of SiO_2, [Al] in steel would reduce CaO and MgO in slag to supply [Ca] and [Mg] into molten steel. $MgO{\cdot}Al_2O_3$ inclusions

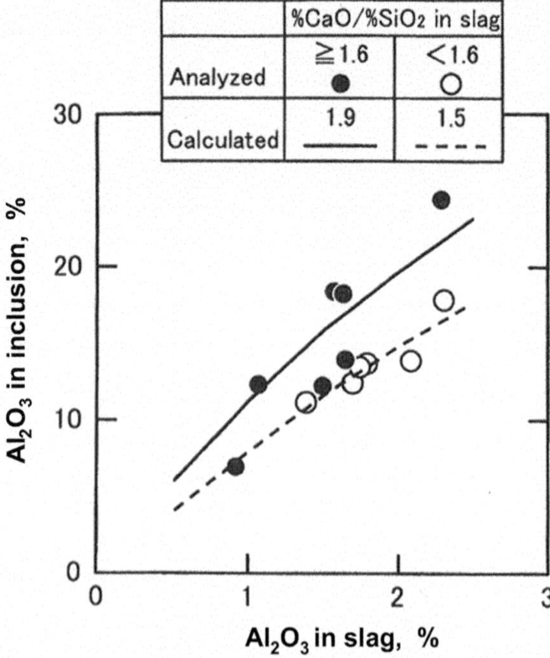

Fig. 1.37 Changes of Al_2O_3 in slag and inclusions, reprinted from Ref. [98], copyright 2007, with permission from ISIJ

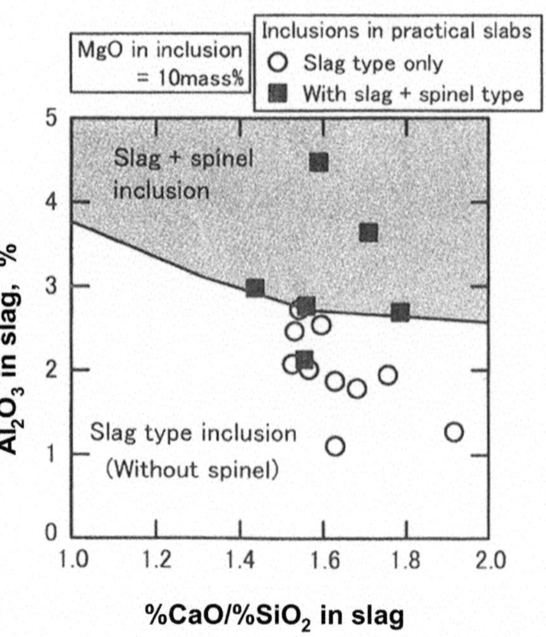

Fig. 1.38 Effect of slag on control of spinel, reprinted from Ref. [98], copyright 2007, with permission from ISIJ

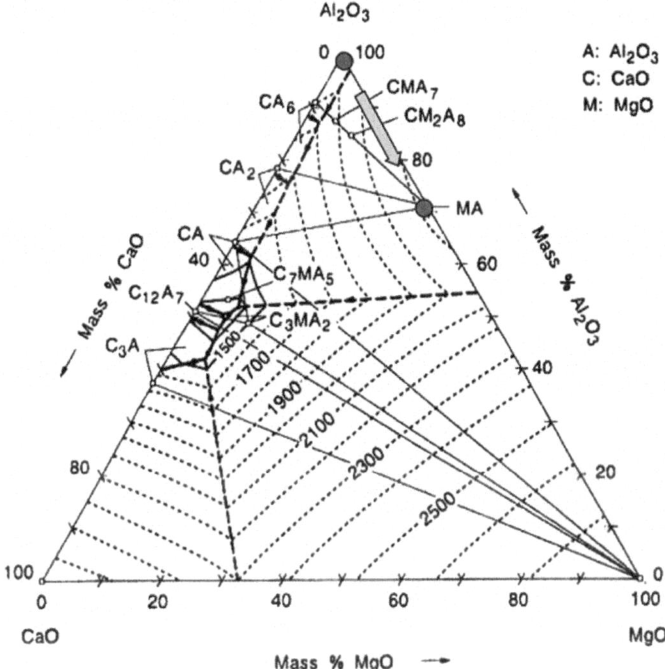

Fig. 1.39 Evolution of inclusions in stainless steel, reprinted from Ref. [99], copyright 2003, with permission from AIST

in steel would be initially transformed into MgO. However, afterwards, the MgO inclusions would be further transformed into calcium magnesium aluminosilicate inclusions by [Ca] in steel, as shown in Fig. 1.39. With the increase of SiO_2 contents in slag, $MgO{\cdot}Al_2O_3$ inclusions became more stable in steel. The reason was that SiO_2 in slag would react with [Al] in molten steel and consume [Al]. As a result, reaction of [Al] in steel and CaO in slag would be weakened. Hence, $MgO{\cdot}Al_2O_3$ inclusions in steel would be more stable, as shown in Figs. 1.40, 1.41 and 1.42. Du et al. reported that $MgO{\cdot}Al_2O_3$ can still be transformed into low-melting-point calcium magnesia aluminate without calcium treatment even [Ca] content in steel was below 0.0001% [101].

From previous studies on inclusions in Al deoxidized steels, it can be known:

(1) During the chemical reaction of slag-steel, spinel inclusions can be easily produced in Al deoxidized steel. However, number of spinel inclusions can be decreased by controlling slag composition and steel compositions. Moreover, if certain [Ca] was contained in steel melts, spinel inclusions can be easily changed into calcium magnesia aluminate inclusions with lower melting point.
(2) Previous studies on spinel inclusions and modification of spinel inclusions mainly focused on the steelmaking of stainless steel, during which the refining slag was featured by low basicity and low alumina content.

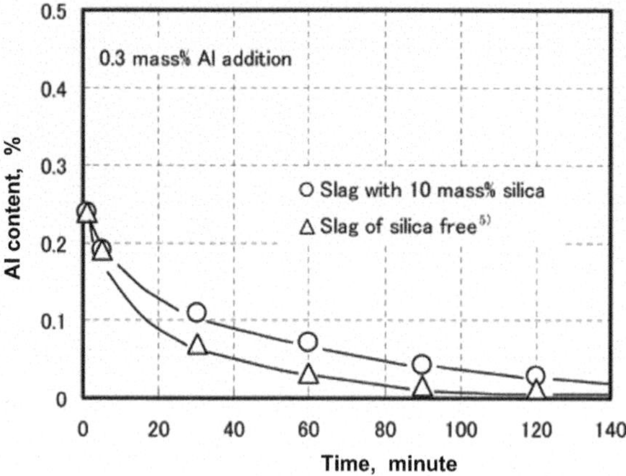

Fig. 1.40 Changes of [Al] in steel, reprinted from Ref. [100], copyright 2004, with permission from ISIJ

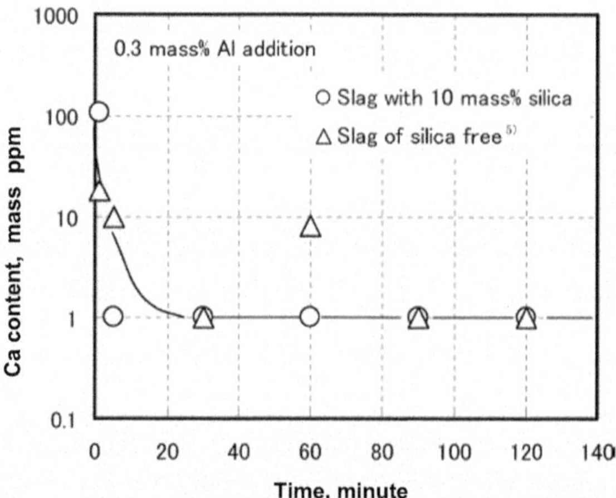

Fig. 1.41 Changes of [Ca] with time, reprinted from Ref. [100], copyright 2004, with permission from ISIJ

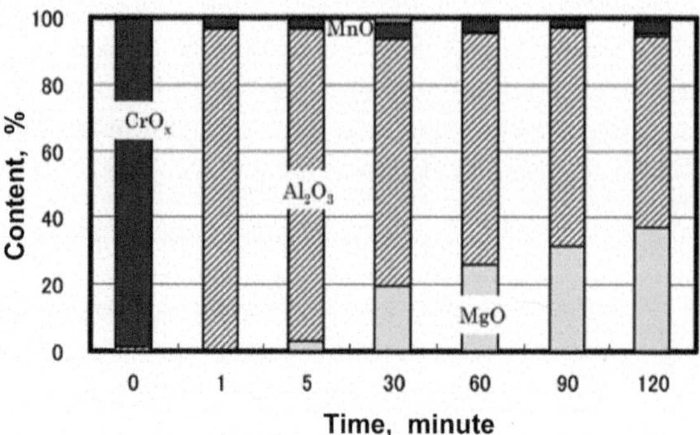

Fig. 1.42 Changes of MgO in inclusions with time, reprinted from Ref. [100], copyright 2004, with permission from ISIJ

(3) Formation mechanism and control of $MgO{\cdot}Al_2O_3$ in high strength low alloy steel were rarely studied, compared to stainless steel. Moreover, steelmaking process of high strength low alloy steel was distinctive from stainless steel. Hence, control and evolution of inclusions in high strength low alloy steel is worthy of deeper investigations.

References

1. Wang, G.D., Liu, X.H., Zhu, F.X., Liu, Z.Y., et al.: Present situation of studying a new generation iron and steel material and its development tendency. Aangang Technology **4**, 1–8 (2005)
2. Weng, Y.Q.: Achievement of "NGSteel" program on China. In: Proceedings of Second International Conference on Advanced Structural Steels, Shanghai (2004)
3. ASTM E 45–5, Standard Test Methods for Determining the Inclusion Content of Steel (2005)
4. GB/T10561–2005–2005/ISO 4967: 1998(E), Steel-Determination of Content of Nonmetallic Inclusions-Micrographic Method Using Standards Diagrams, General Administration of Quality Supervision, Inspection and Quarantine of the People's Republic of China/Standardization Administration of the People's Republic of China (2005)
5. Hilty, D.C.: Inclusions in Steel in Electronic Furnace Steel Making ed. By Taylor C R. Iron and Steel Society. Inc., Bookcrafters Inc., Chelsea, MI (1985)
6. Malkiewica, T., Rudnik, S.: Deformation of non-metallic inclusions during rolling of steel. Journal of Iron and Steel Institute **201**, 33–38 (1963)
7. Rudnik, S.: Discontinuities in hot-rolled steel caused by non-metallic inclusions. Journal of Iron and Steel Institute **204**(4), 374–376 (1966)
8. Atkinson, H.V., Shi, G.: Characterization of inclusions in the steel: a review including the statistics of extreme method progress in material. Prog. Mater Sci. **48**(5), 457–520 (2002)
9. Kiessling, R., Lauge, N.: Nonmetallic inclusions in steel. **5**, 251 (1980)
10. Bernard, G., Ribound, P.V., Urbain, G.: Oxide inclusions' plasticity. La Revue de Metallurgie-CIT **78**(5), 421–433 (1981)
11. Cogne, J.Y.: Cleanness and fatigue life of bearing steels. Clean Steel **3**, 21 (1986)
12. Cogne, J.Y., Heritier, B., Monnot, J.: Cleanness and fatigue life of bearing steel. Production and application of clean steels. Journal of Iron and Steel Institute **6**, 26–31 (1970)
13. Ders, J., Thivard, B., Genta, A.: Improvement of steel wire for cold heading. Wire Journal International **10**, 73–76 (1992)
14. Bertrand, C., Molinero, J., Landa, S., Elvira, R., et al.: Metallurgy of plastic inclusions to improve fatigue life of engineering steels. Ironmaking Steelmaking **30**(2), 165–169 (2003)
15. Brooksbank, D., Andrews, K.W.: Thermal expansion of some inclusions found in steels and relation to tessellated stresses. Journal of Iron and Steel Institute **206**, 595–599 (1968)
16. Brooksbank, D.: Thermal expansion of calcium-aluminate inclusions and relation to tessellated stresses. Journal of Iron and Steel Institute **208**, 495–499 (1970)
17. Brooksbank, D., Andrews, K.W.: Stress field around inclusions and their relation to mechanical properties. Production and application of clean steels, Balatonfured, Hungary. Journal of Iron and Steel Institute June, 186–198 (1970)
18. Brooksbank, D., Andrews, K.W.: Stress field around inclusions and their relation to mechanical properties. Journal of Iron and Steel Institute 246–255 (1972)
19. Kiessling, R.: Non-Metallic Inclusion in Steel, 93, 46. The Institute of Metals, London (1989)
20. Kiessling, R., Nordberg, H.: Inference of inclusions on mechanical properties of steel. Production and Application of Clean Steels, Balatonfured, Hungary, the Iron & Steel Institute 179–185 (1970)

21. Kiessling, R., Nordberg, H.: Influence of inclusions on mechanical properties of steel: critical inclusion size. Clean Steel, Swedish Contribution IVA **1**, 159–169 (1971)
22. Jiang, G.C.: Clean Steel and Secondary Refining, pp. 8–9. Shanghai Scientific & Technical Publishers (1995) (in Chinese)
23. Zhang, S.H.: Alloy Steel, pp. 307, 316. Metallurgical Industry Press, Beijing (1981) (in Chinese)
24. Duckworth, W.E., Ineson, E.: Clean steel 1, p. 87. London: Iron and Steel Institute (1963)
25. Murakami, Y.: Quantitative evaluation of effects of defects and non-metallic inclusions on fatigue strength of metals. Tetsu-to-Hagané **75**(8), 1267–1277 (1989)
26. Tomita, Y.: Improved mechanical properties of ultrahigh strength 0.4C-Cr-Mo-Ni steel through modification of sulphide inclusion shape and microstructure control. Materials Science and Technology **5**, 1084–1089 (1989)
27. Murakami, Y., Kodama, S., Konuma, S.: Quantitative evaluation of effect of non-metallic inclusions on fatigue strength of high strength steel. Int. J. Fatigue **11**(5), 291 (1989)
28. Murakami, Y., Kodama, S., Konuma, S.: Basic fatigue mechanism and evaluation of correlation between the fatigue fracture stress and the size and location of non-metallic inclusions. Int. J. Fatigue **11**(5), 299 (1989)
29. Melander, A., Larsson, M.: The effect of stress amplitude on the cause of fatigue crack initiation in a spring steel. Int. J. Fatigue **15**(2), 119–131 (1993)
30. Zhang, D.T.: Identification of nonmetallic inclusion in steel, vol. 84, pp. 255, 288. National Defense Industry Press, Beijing (in Chinese)
31. Yin, R.Y.: Progress in steel quality-special steel, pp. 99, 101. Metallurgy Industry Press, Beijing (1995) (in Chinese)
32. Tamura, F., Suito, H.: Thermodynamics of oxygen and nitrogen in liquid iron equilibrated with CaO-SiO_2-Al_2O_3 slags. Metall. and Mater. Trans. B. **24B**, 121–130 (1993)
33. Kimura, T., Suito, H.: Calcium deoxidation equilibrium in liquid iron. Metall. and Mater. Trans. B. **25B**, 33–42 (1994)
34. Lee, K.R., Suito, H.: Activities of Fe_tO in CaO-SiO_2-Al_2O_3-Fe_tO (<5 pct) slags saturated with liquid iron. Metall. and Mater. Trans. B. **25B**, 893–902 (1994)
35. Ohta, H., Suito, H.: Activities of SiO_2 and Al_2O_3 and activity coefficients of Fe_tO and MnO in CaO-SiO_2-Al_2O_3-MgO slags. Metall. and Mater. Trans. B. **29B**, 119–129 (1998)
36. Park, J.H., Kim, D.S.: Effect of CaO-Al_2O_3-MgO slag on the formation of MgO-Al_2O_3 inclusions in ferritic stainless steel. Metallurgical and Materials Transactions B **36B**(4), 495–501 (2005)
37. Tang, S.G.: Influence of composition of refining slag in LF-VD process on metallurgical effects. Steelmaking **17**(4), 29–32 (2001) (In Chinese)
38. Zhang, L.F., Thomas, B.G.: State of art in evaluation and control of steel cleanliness. ISIJ Int. **43**(3), 271–291 (2003)
39. Yan, X.C., Zhou, X.J., Liu, Y.J.: Production of Al-killed clean steel with 150t ladle furnace. Southern Metals **5**, 1–4 (2002). (in Chinese)
40. Zhang, X.S., Guan, Y., Lv, C.F., et al.: Development of new type refining slag for LF and its application. Angang Technology **2**, 33–37 (2006). (in Chinese)
41. Lin, G.W.: Function and compounding of slag for refining in ladle furnace. Special Steel **22**(6), 28–29 (2001) (in Chinese)
42. Jiang, Z.H., Li, Y., Jiang, M.F., et al.: Effect of the refining slag composition on the inclusions in steel. In: Proceedings of the 4th CSM Steel Congress, pp. 222–226. Metallurgical Industry Press, Beijing (2003)
43. Yoon, B.H., Heo, K.H., Kim, J.S., Sohn, H.S.: Improvement of steel cleanliness by controlling slag composition. Ironmaking Steelmaking **29**(3), 215–218 (2002)
44. Hino, M., Kitagawa, S., et al.: Sulphide capacities slags of CaO-Al_2O_3-MgO and CaO-Al_2O_3-SiO_2. ISIJ Int. **33**(1), 36–42 (1993)
45. Ohta, M., Kubo, T., Morita, K.: Effects of CaF_2, MgO and SiO_2 addition on sulphide capacities of the CaO-Al_2O_3 slag. Tetsu-to-Hagané **89**(7), 8–15 (2003)

46. Zhong, S.S., Wang, C.S.: Bearing Steel. Metallurgical Industry Press, Beijing (2000).(in Chinese)
47. Ke, S.H.: A summary on desulfurization test of steel by synthetic refining slag. Steelmaking **17**(3), 10–15 (1991) (in Chinese)
48. Wan, Z.Y., Guo, S.X.: Industrial test of synthetic solid slag for desulfurization in LF. Iron and Steel **30**(9), 14–18 (1995) (in Chinese)
49. Wu, G.S.: Sulphur absorption rate of some pre-melted fluxes during heating period. Ironmaking Steelmaking **14**(6), 291 (1987)
50. Ogura, Y., Kikuchi, Y., Hasegawa, T., et al.: Development of secondary refining process and its application to production of clean steel. Tetsu-to-Hagané **72**(9), 1309–1315 (1986)
51. Shibata, H., Emi, T.: In-situ observation of metals and non-metallic inclusions and precipitates by confocal scanning laser microscope. The Bulletin of the Japan Institute Journal of Metals **36**(8), 809–813 (1997)
52. Chikama, H., Shibata, H., Suzuki, M., et al.: In-situ observation of solidification by using a control laser microscope. CAMP-ISIJ **8**, 1128 (1995)
53. Chikama, H., Shibata, H., Emi, T., et al.: In-situ real time observation of planar to cellular and cellular to dendritic transition of crystals growing in Fe-C Alloy melts. Mater. Trans. JIM **37**(4), 620–626 (1996)
54. Chikama, H., Yu, C., Shibata, H., et al.: Application of confocal laser microscope for material technology. Bulletin of IAMP, Tohoku University **51**, 35–40 (1995)
55. Yin, H.B., Emi, T., Chikama, H.: Determination of free Energy of δ-ferrite/γ austenite interphase boundary of low carbon steels by in-situ observation. ISIJ Int. **38**(8), 794–801 (1998)
56. Chikama, H., Yin, H.B., Suzuki, M., et al.: In-situ observation of engulfment and pushing of inclusions by advancing liquid/solid interface in steel melt. CAMP-ISIJ **9**, 596–597 (1996)
57. Yin, H.B., Chikama, H., Emi, T., et al.: In-situ observation of collision agglomeration and cluster formation of alumina inclusion particles on steel melts. ISIJ Int. **37**(10), 936–945 (1997)
58. Yin, H.B., Chikama, H., Emi, T., et al.: Characteristics of agglomeration of various inclusion particles on molten steel surface. ISIJ Int. **37**(10), 946–955 (1997)
59. Chikama, H., Yin, H.B., Emi, T.: The capillary effect promoting collision and agglomeration of inclusion particles at the inert gas-steel interface. Philosophical Transactions of the Royal Society, London, A **356**, 957–966 (1998)
60. Yin, H.B., Chikama, H., Emi, T., et al.: Capillary attraction of solid particles at gas/steel melt interface. In: Eustathopoulos, N., Sobezak, N. (eds.) Proceedings of International Conference on High Temperature Capillary. Cracow, Poland (1997)
61. Bouris, D., Bergeles, G.: Investigation of inclusion re-entrainment from the steel-slag interface. Metallurgical and Materials Transaction B **29B**, 641–664 (1998)
62. Cramb, A.W., Jimbo, I.: Ladle metallurgy: fundamentals of the slag/metal interface. Scaninject VII, Luleo **20**(1), 89–109 (1995)
63. Kozakevitch, P., Lucas, L.: The part played by surface phenomena in the elimination of solid inclusions in a melt bath. Revue de Metallurgie **65**, 589–598 (1996)
64. Kozakevitch, P., Olette, M.: Role of surface phenomena in the mechanism for eliminating solid inclusions. Revue de Metallurgie **68**(10), 636–646 (1971)
65. Kozakevitch, P., Olette, M.: Role of surface phenomena in the mechanism of removal of solid inclusions. In: Production and Application of Clean Steel International Conference, Balatonfured, Hungary, pp. 42–49. The Iron and Steel Institute, London (1970)
66. Vriboud, P., Olette, M.: Mechanisms of some of the reactions involved in secondary refining. Iron and Steel Institute of Japan **22**(8), 879–889 (1982)
67. Nakajima, K., Okamura, K.: Proceeding of 4th International Conference on Molten Slags and Fluxes, pp. 505. ISIJ, Tokyo (1992)
68. Strandh, J., Nakajima, K., Eriksson, R., et al.: Solid inclusion transfer at a steel-slag interface with focus on tundish conditions. ISIJ Int. **45**(11), 1597–1606 (2005)

69. Choi, J.Y., Lee, H.G., Kim, J.S.: Dissolution rate of Al_2O_3 into molten $CaO-SiO_2-Al_2O_3$ slags. ISIJ Int. **42**(8), 852–860 (2002)
70. Choi, J.Y., Lee, H.G.: Wetting of solid Al_2O_3 with molten $CaO-Al_2O_3-SiO_2$. ISIJ Int. **42**(9), 1348–1355 (2003)
71. Martin, V., George, S., Seetharaman, S.: The ability of slags to absorb solid oxide inclusions. ISIJ Int. **46**(3), 450–457 (2006)
72. Seetharaman, S., Du, S.H.: Viscosities of high temperature systems: a modeling approach. ISIJ Int. **37**(2), 109–118 (1997)
73. Kawai, Y., Shiraishi, S. (eds.): Handbook of Physico-chemical Properties at High Temperature, pp. 130, 195. ISIJ, Tokyo (1988)
74. Suito, H., Inoue, R., et al.: Aluminum-oxygen equilibrium between $CaO-Al_2O_3$ melts and liquid iron. ISIJ Int. **31**(12), 1381–1388 (1991)
75. Inoue, R., Suito, H.: Thermodynamics of O, N, and S in liquid Fe equilibrated with $CaO-Al_2O_3-MgO$ slags. Metall. and Mater. Trans. B. **25B**, 235–244 (1994)
76. Ohta, H., Suito, H.: Activities in $MnO-SiO_2-Al_2O_3$ slags and deoxidation equilibria of Mn and Si. Metall. and Mater. Trans. B. **27**, 263–270 (1996)
77. Ohta, H., Suito, H.: Activities in $CaO-MgO-Al_2O_3$ slags and deoxidation equilibria of Al. Mg and Ca. ISIJ Int. **36**(8), 983–990 (1996)
78. Ohta, H., Suito, H.: Deoxidation equilibria of calcium and magnesium in liquid iron. Metall. and Mater. Trans. B. **28B**, 1131–1139 (1997)
79. Suito, H., Inoue, R.: Thermodynamics on control of inclusions composition in ultra-clean steels. ISIJ Int. **36**(5), 528–536 (1996)
80. Shiwaka, K., Koarai, J., et al.: Super clean steel for valve spring. R&D Technical Report of Kobelco **35**(4), 79–82 (1985)
81. Faulring, G.M., Medina, G.: Chaparral's steelmaking practice—then, now and why. In: Proceedings of ISS Electric Furnace Conference, Nashville, vol. **52**, pp. 219–229 (1994)
82. Pielet, H.M., Bhattacharya, D.: Thermodynamics of nozzle blockage in continuous casting of calcium-containing steels. Metall. Trans. B **15**(3), 547–562 (1984)
83. Tönshoff, H.K., Kaestner, W., Schnadt, R.: Verbesserung der Dauerhaltbarkeit von Ventilfederdraht. Stahl Eisen **109**(21), 1011–1015 (1989)
84. Larsen, K., Fruehan, R.J.: Calcium modification of oxide inclusions. Trans. ISS **17**, 45–52 (1990)
85. Bannenberg, N.: Inclusion modification to prevent nozzle clogging. In: Steelmaking Conference Proceedings, pp. 57–463 (1995)
86. Ye, G., Josson, P., Lund, T.: Thermodynamics and kinetics of the modification of Al_2O_3 inclusions. ISIJ Int. **36**(Supplement), S105–S108 (1996)
87. Tamamoto, S., Sasaki, K., Nashiwa, H., et al.: Alloying by shooting alloy bullets into molten steel in the ladle. Tetsu-to-Hagané **73**(13), 2110–2125 (1977)
88. Cicutti, C., Vladez, M., Perez, T, et al.: Optimization of calcium treatment to improve castability. In: 84th Steelmaking Conference Proceedings, pp. 871–882, Baltimore (2001)
89. Okuyama, G.: Effect of slag compositions on formation rate of spinel inclusions in stainless steel melt. Curr. Adv. Mater. Process. **10**, 847–847 (1997)
90. Todoroki, H., Mizuno, K., Noda, M., Tohge, T.: Formation mechanism of spinel inclusions in 304 stainless steel deoxidized with ferrosilicon alloys. In: Proceedings of 2001 Steelmaking Conference, pp. 331–341 (2001)
91. Itoh, H., Hino, M., Ban-ya, S.: Thermodynamics on the formation of spinel non-metallic inclusion in liquid steel. Metall. and Mater. Trans. B. **28B**, 953–956 (1997)
92. Seo, W.G., Han, W.H., Kim, J.S., et al.: Deoxidation equilibria among Mg, Al and O in liquid iron in the presence of $MgO{\cdot}Al_2O_3$ Spinel. ISIJ Int. **43**(2), 201–208 (2003)
93. Kim, J.W., Kim, S.K., et al.: Formation mechanism of Ca-Si-Al-Mg-Ti-O inclusions in type 304 stainless steel. ISIJ Int. **35**(Supplement), S140–S143 (1996)
94. Cha, W.Y., Kim, D.S., Lee, Y.D., et al.: A thermodynamic study on the inclusions formation in ferritic stainless steel melt. ISIJ International **44**(7), 1134–1139 (2004)

95. NiShi, T., Shinme, K.: Formation of spinel inclusions in molten stainless steel under Al deoxidization with slags. Tetsu-to-Hagané **84**(12), 837–843 (1998)
96. Park, J.H.: Formation mechanism of spinel-type inclusions in high-alloyed stainless steel melts. Metall. and Mater. Trans. B. **38B**, 657–663 (2007)
97. Ehara, Y., Yokoyama, S., Kawakami, M.: Formation mechanism of inclusions containing $MgO{\cdot}Al_2O_3$ spinel in type-304 stainless steel. Tetsu-to-Hagané **93**(3), 208–214 (2007)
98. Ehara, Y., Yokoyama, S., Kawakami, M.: Control of formation of spinel inclusion in type 304 stainless steel by slag composition. Tetsu-to-Hagané **93**(7), 475–482 (2007)
99. Todoroki, H., Mizuno, K.: Variation of inclusions composition in 304 stainless steel deoxidized with aluminum. Iron and Steelmaker **30**(3), 60–67 (2003)
100. Todoroki, H., Mizuno, K.: Effect of silica in slag on inclusion compositions in 304 stainless steel deoxidized with aluminum. ISIJ Int. **44**(8), 1350–1357 (2004)
101. Kang, Y.J., Li, F., Morita, K., Du, S.C.: Mechanism study on the formation of liquid calcium aluminate inclusion from MgO-Al_2O_3 spinel. Steel Res. Int. **77**(11), 785–792 (2006)

Chapter 2
Formation of Spinel Inclusions in HSLA Steel Refined by $CaO–SiO_2–Al_2O_3–MgO$ Slag

Abstract In this chapter, formation of spinel in high strength low alloy steel (HSLA) is discussed, as spinel has been noticed as a typical type of inclusions causing fatigue problems. The commercial steel grade 42CrMo was chosen in the laboratorial slag-steel experiments, in which compositions of the $CaO–SiO_2–Al_2O_3–MgO$ refining slag were changed to elucidate the influences of refining slags on the formation of spinel inclusions. The results indicated important influence of refining slag on the contents of [Mg] and [Al] in liquid steel, which in turn affect the formation of spinel inclusions.

As introduced intensively in previous chapter, spinel ($MgO{\cdot}Al_2O_3$) inclusions can be easily produced in Al deoxidized steel during the refining process. In the past, spinel was intensively discussed for stainless steel while rarely for the low alloy steel. Because of poor deformability, spinel inclusions like to adhere to the head of stoppers or inner wall of nozzles to deteriorate castability of molten steel. During continuous casting process, the clogs can be washed down by the pouring stream of molten steel as very large inclusions to threaten anti-fatigue property of steel. Therefore, detrimental influence of inclusions on ultra-high cycle fatigue life of steels aroused great attention in recent years, and how to prevent high-melting-point inclusions is of great significance in steelmaking.

In this chapter, formation of spinel inclusions in high strength low alloy steel was discussed by a series of laboratory scale slag-steel reaction experiments. The steel melts were refined by the $CaO–SiO_2–Al_2O_3–MgO$ slag, in which slag compositions were changed to elucidate the effects of slag compositions on inclusions. In the experiments, soluble aluminum contents in steel melts were within 0.01 mass%.

Contents of this chapter was a reprint of the authors publication on ISIJ International, 2008, 48(7): 885–890, copyright 2008, with permission from ISIJ

M. Jiang and X. Wang, *Slag-Steel Reaction and Control of Inclusions in Al Deoxidized Special Steel*, Engineering Materials, https://doi.org/10.1007/978-981-19-3463-6_2

2.1 Slag-Steel Reaction Experiments in Lab

An electrical resistance furnace with heating bars of Si–Mo was used to carry out the slag-steel reaction experiments. During the experiments, steel sample (150 g, chemical compositions are listed in Table 2.1) and slag (30 g) were put and melted together in MgO crucibles set at the even temperature zone of the reaction tube at 1873 K. The reaction tube was sealed at the two ends and evacuated by deoxidized flowing argon gas all the time. When the temperature was elevated to 1873 K, the steel and slag were held for a determined period of time confirmed by pre-equilibrium experiments. Afterwards, the MgO crucibles with steel were taken out and rapidly quenched in water.

2.1.1 Determination of Reaction Time by Pre-equilibrium Experiments

Before equilibrium experiments, pre-experiments have been carried out to find the holding time for attainment of equilibrium between slag and molten steel, during which the crucible containing slag and steel was held at 1873 K for different time. Then, the crucible was pulled out with rapid quenching in water. And the obtained steel and slag samples were prepared for chemical composition analysis.

2.1.2 Slag-Steel Reaction Experiments

After the time was determined, slag-steel reaction experiments were conducted at 1873 K, during which steel melts were without stirring. The crucible was then taken out and quenched rapidly in water. The obtained steel and slag samples were prepared for chemical composition analysis. Specimens of steel were also prepared for inclusion observation. Thirty inclusions on the cross-sectioned plane of the sample were randomly inspected and analyzed by SEM–EDS to get the overall and statistical information of inclusions such as morphology, size, chemical compositions and so on. The chemical compositions of slag samples were analyzed by an X-ray fluorescence spectrometer.

Table 2.1 Chemical compositions of 42CrMo steel material (mass%)

C	Si	Mn	P	S	Cr	Mo
0.42	0.25	0.65	0.01	0.002	1.15	0.2

2.2 Results and Discussion

2.2.1 Determined Reaction Time Between Steel and Slag

Chemical compositions of steel and slag samples obtained in the pre-equilibrium experiments were listed in Table 2.2.

It was found that after holding for 90 min, total oxygen contents in the melt changed very little. The result indicated that 90 min were sufficient for the floatation of inclusions in molten steel up to slag. While aluminum contents in steel melts changed little after the melts were held for 60 min, as shown in Fig. 2.1. It could be concluded that 90 min are sufficient to attain chemical reaction equilibrium between slag and steel.

Table 2.2 Chemical compositions of melts and slags in pre-equilibrium experiments (mass%)

Exps (min)	Slag				Metal	
	CaO	SiO_2	Al_2O_3	MgO	[Al]	T[O]
30	55.89	11.01	16.22	2.25	0.0079	0.0021
60	60.96	13.06	18.6	4.2	0.0050	0.0016
90	61.82	12.86	18.71	4.54	0.0048	0.0010
120	57.64	12.27	20.89	7.88	0.0051	0.0013
180	59.52	12.28	19.39	6.63	0.0050	0.0010

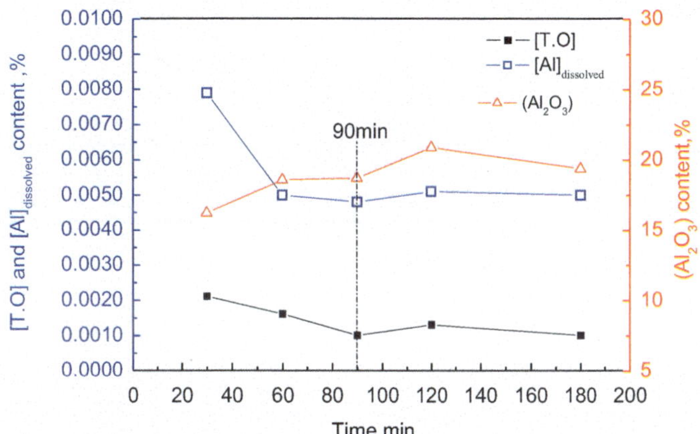

Fig. 2.1 Time for attainment of equilibrium determined by experiments

2.2.2 *Inclusions in Steel Melts*

Compositions of steel and slag samples in equilibrium experiments were listed in Table 2.3.

It was found during the SEM–EDS observation that $MgO \cdot Al_2O_3$ inclusions widely dispersed in steel samples. In particular, they were very small (about 2–4 μm), globular, uniformly distributed particles other than angular particles except in experiment No. 4, as shown in Figs. 2.2, 2.3, 2.4, 2.5, 2.6, 2.7 and 2.8.

Table 2.3 Chemical compositions of slag and steel sample in equilibrium experiments (mass%)

Exp. No.	Slag				Steel		
	CaO	SiO_2	Al_2O_3	MgO	[Al]	[Mg]	$[O]_{calculated}$
No. 1	54.69	18.50	18.74	7.38	0.0045	0.0005	0.00051
No. 2	45.66	15.20	27.58	10.96	0.0054	0.0004	0.00058
No. 3	34.76	11.28	46.01	6.98	0.0060	0.0004	0.00084
No. 4	50.77	10.74	27.38	10.00	0.0053	0.0005	0.00042
No. 5	35.87	7.21	43.85	12.31	0.0073	0.0005	0.00054
No. 6	52.66	8.10	27.22	11.22	0.0084	0.0010	0.00022
No. 7	38.05	5.72	44.71	11.05	0.0100	0.0008	0.00032

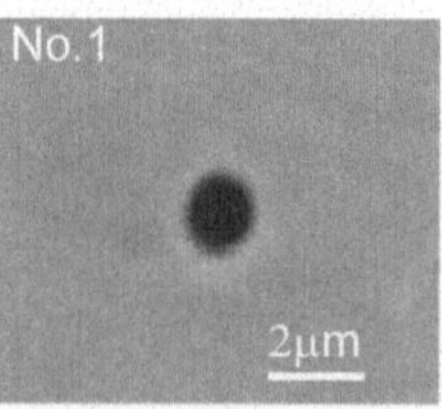

Element	Mg	Al	Si	Cr	Mn	Fe
Atom(%)	10.02	19.25	1.61	0.74	1.00	67.38

Fig. 2.2 Typical inclusion detected in steel sample No. 1

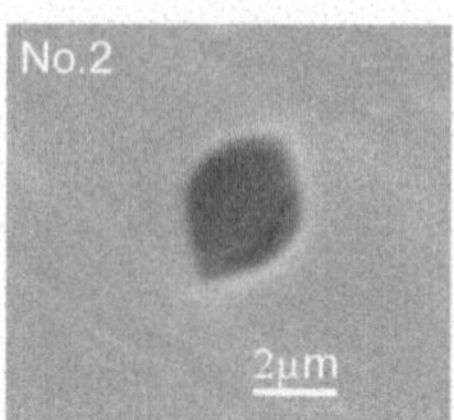

Element	Mg	Al	Si	Cr	Fe
Atom(%)	13.44	28.93	2.96	1.09	53.58

Fig. 2.3 Typical inclusion detected in steel sample No. 2

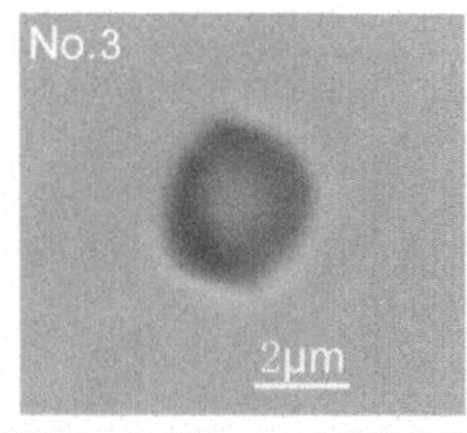

Element	O	Mg	Al	Ca	Cr	Fe
Atom(%)	51.01	11.56	28.95	1.33	0.31	6.84

Fig. 2.4 Typical inclusion detected in steel sample No. 3

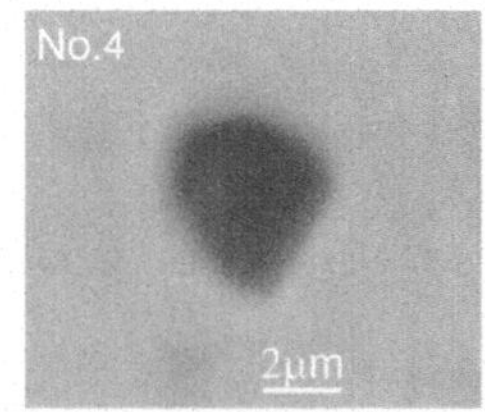

Element	O	Mg	Al	Mn	Fe
Atom(%)	45.00	10.57	26.63	0.97	16.83

Fig. 2.5 Typical inclusion detected in steel sample No. 4

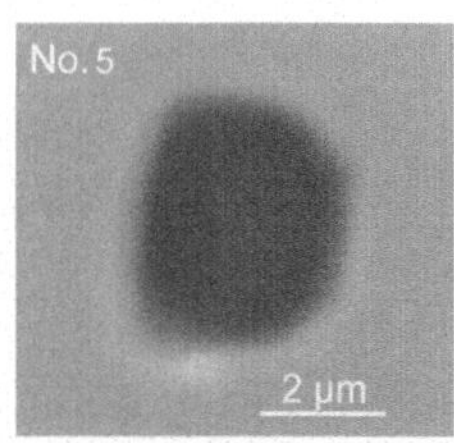

Element	O	Mg	Al	Fe
Atom(%)	52.34	13.69	27.81	6.16

Fig. 2.6 Typical inclusion detected in steel sample No. 5

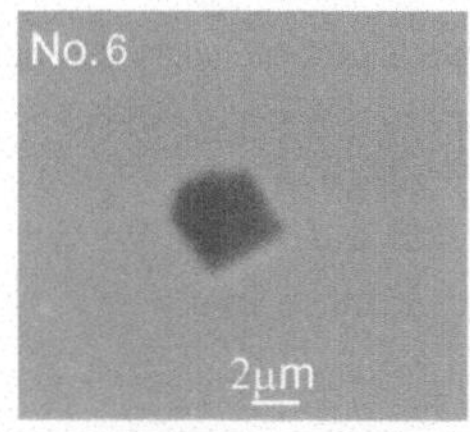

Element	O	Mg	Fe
Atom(%)	37.04	47.50	15.46

Fig. 2.7 Typical inclusion detected in steel sample No. 6

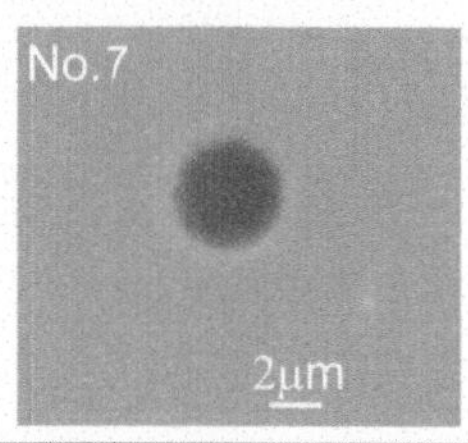

Element	O	Mg	Al	Si	Ca	Fe
Atom(%)	41.11	29.46	12.91	0.61	2.05	13.86

Fig. 2.8 Typical inclusion detected in steel sample No. 7

Magnesia content in spinel inclusions increased with the rise of magnesium contents in steel melts. Particularly, many rectangular MgO particles existed in the steel sample No. 6, in which magnesium content was 0.0010% and highest among all steel samples. The reason would be discussed in details in the following part. It was noticeable that shapes of spinel inclusions were modified from spherical to sharp angular with the increase of magnesium content in molten steel. While trace amount of calcium in inclusions was helpful to modify the inclusions into spherical shape, as indicated in Figs. 2.2, 2.3, 2.4, 2.5, 2.6, 2.7 and 2.8.

2.2.3 [Al]–[O] Deoxidation Equilibrium

When chemical reaction equilibrium was established between slag and steel, contents of aluminum and magnesium in steel were in equilibrium with slag, as expressed by the following reaction [1, 2].

$$(Al_2O_3) = 2[Al] + 3[O] \tag{2.1}$$

$$\log K_{Al} = a^2_{[Al]} \cdot a^3_{[O]}/a_{Al_2O_3} = 20.57 - 64{,}000/T \tag{2.2}$$

where K_i was the equilibrium constant, $a_{[i]}$ was the activity of component i in steel, which can be calculated by the standard state of infinite dilute solution, and a_j was the activity of component j in slag, whose standard state was pure solid. The activities of [Al], [O] can be calculated with the following Eqs. (2.3) and (2.4),

$$\log a_{[Al]} = \sum e^j_{Al}[\text{mass\%}j] + \log[\text{mass\%Al}] \tag{2.3}$$

$$\log a_{[O]} = \sum e^j_{O}[\text{mass\%}j] + \log[\text{mass\%}O] \tag{2.4}$$

where $e^i{}_j$ was the first order interaction coefficient between elements i and j, which were given in Table 2.4. Suito et al. deduced a regression equation to calculate the

Table 2.4 Interaction coefficients used in thermodynamic calculations at 1873 K in present works [4]

	C	Si	Mn	P	S	Cr	Mo	Al	O
Al	0.091	0.0056	–	0.033	0.030	0.012	–	0.045	– 6.6
O	– 0.436	– 0.131	– 0.021	0.07	– 0.133	– 0.04	0.0035	– 3.9	– 0.2
Mg	– 0.15 [5]	– 0.09 [6]	–	–	– 1.38	0.05	–	– 0.12	– 430

activity of alumina in CaO–SiO_2–Al_2O_3–MgO slag at 1873 K as follows [3]:

$$\log a_{Al_2O_3} = \{-0.275(\%CaO) + 0.167(\%MgO)\}/(\%SiO_2) + 0.033(\%Al_2O_3) - 1.560 \quad (2.5)$$

Substituting the activity of Al_2O_3 estimated by Eq. (2.5) using the slag compositions given in Table 2.3, Eqs. (2.3) and (2.4) to Eq. (2.2), dissolved oxygen content can be calculated. The oxygen contents in the molten steel in equilibrium with alumina in slag phase at 1873 K were obtained and given in Table 2.3.

2.2.4 Saturation Behavior of MgO in Slag

Ohta et al. plotted the iso-solubility lines of MgO in (CaO + MgO)–SiO_2–Al_2O_3 quasi-ternary system, as shown in Fig. 2.9 [3]. The heavy lines represented the boundaries of different phases, and the light lines represented the iso-solubility lines of MgO. It can be seen from the iso-solubility lines of MgO that, with the increase in Al_2O_3 content at a given SiO_2 content in slag, solubility of MgO in MgO-saturated region increased towards the boundary of MgO–MgO·Al_2O_3 saturation region and thereafter, decreased in the MgO·Al_2O_3-saturated region.

Compositions of slag samples listed in Table 2.2 were projected to this pseudo-ternary system, as indicated by the triangles in Fig. 2.9. Slag Compositions of experiments No. 1 and No. 3 located near the saturation boundary of MgO and C_2S and in the MgO·Al_2O_3 saturated region, respectively, where the MgO solubility was between 5 and 10%. Because MgO contents in slag of experiments No. 1 and No. 3 were 7.38% and 6.98%, respectively, so activity of MgO in the slag of experiment No. 1 and No. 3 can be considered as unity. In the experiments No. 5 and No. 7, slag compositions located near the phase boundary of MgO–MgO·Al_2O_3 with MgO solubility about 10%–15%. Because MgO contents in the slag of experiments No. 5 and No. 7 were between 11.5 and 12.31%, activities of MgO in the two slags can be considered as unity, too. As far as other slags concerned, MgO contents were higher than their solubility, so the activities of MgO in these slags can also be considered as unity.

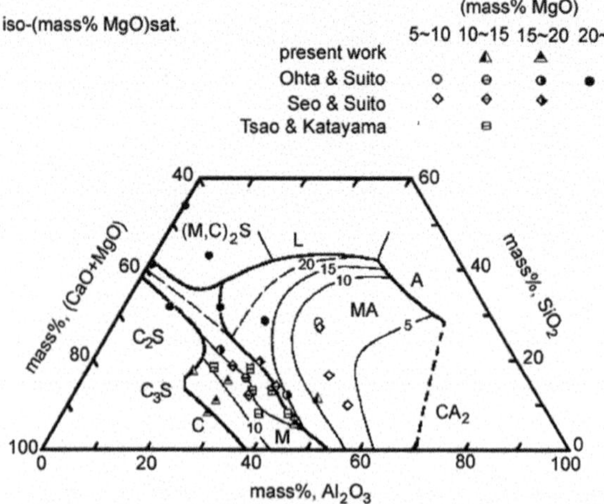

Fig. 2.9 Saturation behavior of MgO in the (CaO + MgO)–SiO_2–Al_2O_3 pseudo-ternary diagram (referred from the Ref. [3], with experimental data of present authors), reprinted by permission from springer nature, Metallurgical and Materials Transactions B, Activities of SiO_2 and Al_2O_3 and activity coefficients of Fe_tO and MnO in CaO–SiO_2–Al_2O_3–MgO slags, Hiroki Ohta and Hideaki Suito, copyright 1998

2.2.5 Chemical Reaction Equilibrium Among Slag-Steel-Spinel Inclusion

According to Suito and Inoue, [4] compositions of inclusions can be estimated from thermodynamic calculation, providing that chemical reaction equilibrium between steel and inclusion had been established. Theoretically, compositions of inclusions would be identical to slag if chemical reaction equilibrium can be completely established among slag, steel and inclusions. Whereas, it is impossible to achieve in the practice. Okuyama et al. proved that chemical reactions between inclusions and steel were fast enough to establish local equilibrium between them [7].

In the present experiments, the equilibria between inclusions and molten steel can be expressed by Eqs. (2.6) to (2.9) [8].

$$(\mathrm{MgO})_{\mathrm{inclusion}} = [\mathrm{Mg}] + [\mathrm{O}] \tag{2.6}$$

$$(a_{\mathrm{MgO}})_{\mathrm{inclusion}} = a_{[\mathrm{Mg}]} \cdot a_{[\mathrm{O}]}/K_{\mathrm{Mg}} \tag{2.7}$$

$$\log a_{[\mathrm{Mg}]} = \sum e^{j}_{\mathrm{Mg}}[\mathrm{mass}\%j] + \log[\mathrm{mass}\%\mathrm{Mg}] \tag{2.8}$$

$$(\mathrm{Al_2O_3})_{\mathrm{inclusion}} = 2[\mathrm{Al}] + 3[\mathrm{O}] \tag{2.9}$$

$$(a_{Al_2O_3})_{inclusion} = a_{[Al]}^2 \cdot a_{[O]}^3 / K_{Al} \quad (2.10)$$

From Eqs. (2.7) and (2.10), it can be deduced that compositions of inclusions can be expressed as a function of the activities of magnesium, aluminum, dissolved oxygen, activity coefficients of MgO and Al_2O_3 and temperature as follows:

$$\log(X_{MgO}/X_{Al_2O_3}) = \log[a_{[Mg]}/(a_{[Al]}^2 \cdot a_{[O]}^2)] - \log(\gamma_{MgO}/\gamma_{Al_2O_3}) + \log(K_{Al}/K_{Mg}) \quad (2.11)$$

In Eq. (2.11), X_i and γ_i were the mole fraction and the activity coefficient of element i in inclusions. Hence, logarithm of the concentration ratios of MgO to Al_2O_3 in inclusions, $log(X_{MgO}/X_{Al_2O_3})$ was expected to exhibit linear relation to $log[a_{[Mg]}/(a_{[Al]}^2 \cdot a_{[O]}^2)]$ with a slope of unity at a fixed temperature when the activity coefficient term in Eq. (2.11) is not significantly affected by the chemical compositions of steel.

The relation between observed $\log[a_{[Mg]}/(a_{[Al]}^2 \cdot a_{[O]}^2)]$ and observed $\log(X_{MgO}/X_{Al_2O_3})$ at 1873 K was given in Fig. 2.10 as the scattered solid circles. The dotted line was the fitted line of the experimental data points at 1873 K obtained by a linear regression analysis. It is found that observed $\log(X_{MgO}/X_{Al_2O_3})$ exhibited good linear relation with observed $\log[a_{[Mg]}/(a_{[Al]}^2 \cdot a_{[O]}^2)]$ indicating a slope of about 0.65.

As mentioned above, activity of MgO can be considered as unity. As activity of alumina in slag can be calculated by Eq. (2.10), then the relation between observed $\log(X_{MgO}/X_{Al_2O_3})$ and observed $\log(a_{MgO}/a_{Al_2O_3})$ at 1873 K was plotted and indicated in Fig. 2.11 by solid squares. The dotted line was the fitted line of the experimental data obtained by a linear regression analysis. It could be found that observed $\log(X_{MgO}/X_{Al_2O_3})$ increased with the rise of observed $\log(a_{MgO}/a_{Al_2O_3})$, exhibiting good linear relation with a slope about 0.70.

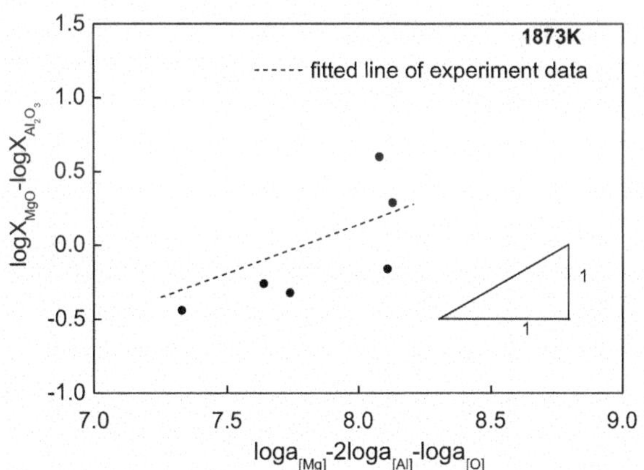

Fig. 2.10 Composition of spinel inclusions, $\log(X_{MgO}/X_{Al_2O_3})$ as a function of $\log[a_{[Mg]}/(a_{[Al]}^2 \cdot a_{[O]}^2)]$ in molten steel at 1873 K

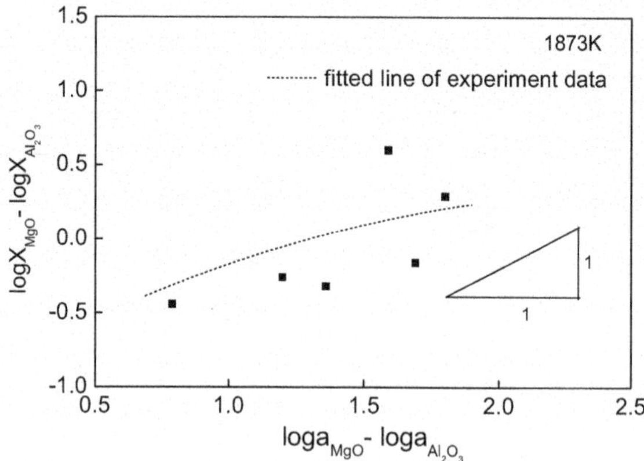

Fig. 2.11 Composition of spinel inclusions, $\log(X_{MgO}/X_{Al_2O_3})$ as a function of $\log(a_{MgO}/a_{Al_2O_3})$ of slag at 1873 K

As inclusions in steel sample No.6 were pure MgO, $X_{Al_2O_3}$ was zero and $\log X_{Al_2O_3}$ would be of non-sense in math. As a result, the numbers of data points in Figs. 2.10 and 2.11were six rather than seven.

Two reasons accounted for scattered data points and why slopes of the lines in Figs. 2.10 and 2.11 were not unity. The reason was that silica in slag influenced greatly the MgO content in inclusion, thus MgO content in spinel in present paper not only depend on the content of MgO in slag but also the content of silica in slag, which rightly caused the scattering of data points and the deviations between the slopes of fitted line and unity.

2.2.6 *Calculation on Stability Phase Diagram of MgO–MgO·Al₂O₃–Al₂O₃*

To explain the formation of inclusions, formation thermodynamic calculations of inclusions were conducted to obtain the stability diagram of MgO, spinel and Al_2O_3. The used equations are listed as follows:

Calculation of MgO/ $MgO{\cdot}Al_2O_3$ boundary [9]

$$4(MgO) + 2[Al] = MgO \cdot Al_2O_3 + 3[Mg] \tag{2.12}$$

$$\log K_{13} = -33.09 + 50{,}880/T \tag{2.13}$$

$$K_{13} = \frac{a_{[Mg]}^3 \cdot a_{MgO\cdot Al_2O_3}}{a_{[Al]}^2 \cdot a_{MgO}^4} \tag{2.14}$$

Calculation of $MgO{\cdot}Al_2O_3/Al_2O_3$ boundary [9]

$$4(Al_2O_3) + 3[Mg] = 3(MgO \cdot Al_2O_3) + 2[Al] \tag{2.15}$$

$$\log K_{16} = 34.37 - 46{,}950/T \tag{2.16}$$

$$K_{16} = \frac{a^2_{[Al]} \cdot a^3{}_{MgO \cdot Al_2O_3}}{a^4_{Al_2O_3} \cdot a^3_{[Mg]}} \tag{2.17}$$

According to Hino et al. [10] when MgO/MgO·Al_2O_3 boundary was calculated at 1873 K, activity of MgO·Al_2O_3 was taken as 0.8 while that of MgO as 0.99 because of very small solubility of Al_2O_3 into MgO. When MgO·Al_2O_3/Al_2O_3 boundary was calculated, activity of MgO–Al_2O_3 was taken as 0.47 while that of Al_2O_3 as unity because of negligibly small solubility of MgO into Al_2O_3. Oxygen concentration was taken as 0.0005%, which was about the average of dissolved oxygen contents steel melts. Compared with magnesium, influence of dissolved oxygen contents on the activity of alumina was ignored because e^O_{Al} was much smaller than e^O_{Mg}. During the calculation of activity coefficients, it was found that influences of dissolved oxygen contents on the activity coefficients of aluminum can be ignored. Substituting Eqs. (2.3), (2.8) to Eqs. (2.14) and (2.17), the phase stability diagram can be calculated, as shown in Fig. 2.12.

It could be found that in the scope of dissolved aluminum of present paper, tiny increases of magnesium contents in molten steel would promote the transformation of MgO·Al_2O_3 and MgO. It was obvious that the Mg and Al contents in molten steel mainly existed in the spinel formation zone, which was the primary reason for the large formation of MgO·Al_2O_3 inclusions. In Experiment No. 6, in which inclusions were mainly rectangular MgO inclusion, the data point (indicated by the bold arrow) positioned rightly at the boundary of MgO/spinel. It meant that MgO can be formed in this experiment. Hence, the calculated diagram agreed with the experimental results.

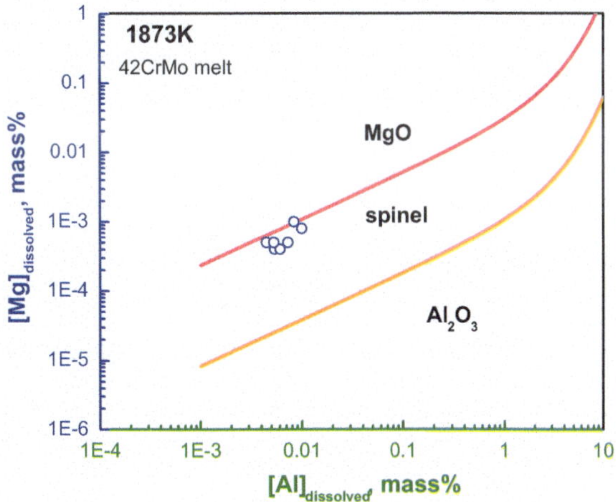

Fig. 2.12 Phase stability diagram of MgO, MgO·Al_2O_3 and Al_2O_3

Itoh et al. evaluated the stable formation region of spinel inclusions in liquid steel by assuming an ideal solid solution behavior of spinel during de-oxidation process [11]. In their work, when aluminum content in molten steel was lower than 0.0010%, MgO inclusions would form directly in molten steel. However, in present work, MgO cannot directly form because there was no magnesium in molten steel at the very beginning of these experiments. All MgO inclusions were evolved from $MgO{\cdot}Al_2O_3$. Moreover, the spinel inclusion region is narrower than that predicted by Itoh et al.

Todoroki et al. evaluated the stable formation region of spinel inclusions in the 304 stainless steel refined by CaO–(SiO_2)–Al_2O_3–MgO–CaF_2 slag [9]. In their work, the shape of spinel inclusion region was similar to that in present paper, but the threshold value for Al_2O_3–$MgO{\cdot}Al_2O_3$ and $MgO{\cdot}Al_2O_3$–MgO transformation was lower than the predictions of the present work. The differences originated from the distinctions in the chemical compositions of steel melts.

According to Todaroki et al. [9] silica in slag would react with soluble aluminum and magnesium, which retarded and even prohibited further transformation of $MgO{\cdot}Al_2O_3$ inclusion to MgO inclusions or CaO–Al_2O_3–MgO system inclusions. In their paper, activity of silica in slag was estimated as about 0.0001 using the regression formula given by Ohta and Suito [4]. In the present paper, activity of silica in slag was calculated about 0.0003 to 0.006 by this regression formula. It was found that activity of silica in slag in experiment No. 6 was about 0.0003, which was the lowest of all the slags. Correspondingly, magnesium content in the steel melt was 0.0010%, which was the highest level among all steel melts. It was noticeable that in experiment No. 6 (aluminum content was 0.084%, and magnesium content was 0.0010%), spinel inclusions were transformed into rectangular MgO particles. Furthermore, it could be found that MgO contents in inclusions increased with the decrease of silica contents in slag, as shown by the EDS results in Figs. 2.2, 2.3, 2.4, 2.5, 2.6, 2.7, 2.8 and Table 2.3.

It can be drawn that in the scope of aluminum contents of the present paper, tiny fluctuations of magnesium contents in steel greatly influenced the transformation of $MgO/MgO{\cdot}Al_2O_3/Al_2O_3$ inclusions in steel. When magnesium contents reached a threshold value, spinel inclusions would be transformed to magnesia inclusions. Silica in slag consumed soluble aluminum especially magnesium in molten steel, thus prohibited the transformation of spinel into MgO inclusions stabilized the existence of $MgO{\cdot}Al_2O_3$ inclusions.

2.3 Summary

Laboratory scale experiments as well as thermodynamic calculations were done to investigate the formation of $MgO{\cdot}Al_2O_3$ inclusions in high strength low alloy strength steel. The following conclusions were obtained:

(1) Pre-equilibrium experiments have been done and it was found that 90 min are sufficient for metal-slag to approach equilibrium;

(2) Inclusions in the low alloy structural steel were mainly $MgO{\cdot}Al_2O_3$ spinel. Contents of MgO in inclusions increased with the rise of Mg contents in molten steel. Spinel inclusions evolved gradually from globular ones into angular ones with the rise of Mg content in molten steel. Trace amount of calcium in inclusions was helpful to modify the angular spinel to globular inclusions in spite of the rise of Mg content in molten steel, which was very positive to improve the fatigue life of the high strength 42CrMo alloyed structural steel.

(3) MgO in slags during experiments in present paper were saturated; thus, activity of MgO could be considered as unity;

(4) Observed $\log(X_{MgO}/X_{Al_2O_3})$ of the inclusions linearly increased with the rise of observed $\log[a_{[Mg]}/(a^2_{[Al]}\cdot a^2_{[O]})]$ in molten steel and with the rise of observed $\log(a_{MgO}/a_{Al_2O_3})$ in slag, exhibiting good linear relation. But the slopes of the fitted lines were not unity. The reason was that silica in slag influenced greatly the MgO content in inclusion, thus the MgO content in spinels depend on not only MgO but also silica content in the slag.

(5) Stability phase diagram of MgO–$MgO{\cdot}Al_2O_3$–Al_2O_3 was obtained. It showed that aluminum and magnesium contents in experiments of present paper mainly positioned in $MgO{\cdot}Al_2O_3$ formation zone, which was the primary reason for the widespread existence of $MgO{\cdot}Al_2O_3$ spinel inclusions in steel. Silica in slag stabilized the existence of $MgO{\cdot}Al_2O_3$ inclusions.

References

1. Todoroki, H., Mizuno, K.: Variation of inclusion composition in 304 stainless steel deoxidized with aluminum. ISS Transactions, Iron and Steelmaker **61**(3), 60–68 (2003)
2. Itoh, H., Hino, M., Ban-Ya, S.: Thermodynamics on the formation of non-metallic inclusion of spinel ($MgO{\cdot}Al_2O_3$) in liquid steel. Tetsu-to-Hagané **84**(2), 85–90 (1998)
3. Ohta, H., Suito, H.: Activities of SiO_2 and Al_2O_3 and activity coefficients of Fe_tO and MnO in CaO-SiO_2-Al_2O_3-MgO slags. Metall. and Mater. Trans. B. **29**(1), 119–129 (1998)
4. Suito, H., Inoue, R.: Thermodynamics on control of inclusions composition in ultraclean steels. ISIJ Int. **36**(5), 528–536 (1996)
5. Sigworth, G.K., Elliott, J.F.: The thermodynamics of liquid dilute iron alloys. Metal Science **8**(1), 298–310 (1974)
6. Ohta, H., Suito, H.: Activities in CaO-MgO-Al_2O_3 slags and deoxidation equilibria of Al, Mg, and Ca. ISIJ Int. **36**(8), 983–990 (1996)
7. Okyuama, G., Yamaguchi, K., Takeuchi, S., et al.: Development of high strength 15Cr ferritic creep resistant steel with addition of tungsten and cobalt. ISIJ Int. **41**, 121–125 (2001)
8. Park, J.H., Kim, D.S.: Effect of CaO-Al_2O_3-MgO slags on the formation of MgO-Al_2O_3 inclusions in ferritic stainless steel. Metall. and Mater. Trans. B. **36B**, 495–502 (2005)
9. Todoroki, H., Mizuno, K.: Effect of silica in slag on inclusion compositions in 304 stainless steel deoxidized with aluminum. ISIJ Int. **44**(8), 1350–1357 (2004)
10. Fujii, K., Nagasaka, T., Hino, M.: Activities of the constituents in spinel solid solution and free energies of formation of MgO, MgO·Al2O3. ISIJ Int., **40**(11), 1059–1066 (2000)
11. Itoh, H., Hino, M., Ya, S.B.: Thermodynamics on the formation of spinel nonmetallic inclusion in liquid steel. Metall. Mater. Trans. B, **28B**(5), 953–956 (1997)

Chapter 3
Formation Thermodynamics of Inclusions in Al Deoxidized Low Alloy Steel

Abstract In this chapter, formation thermodynamics of low-melting-point inclusions in Al-deoxidized steel was conducted. As MgO is widely used for slag making and ladle lining, spinel inclusions can be formed very easily in secondary refining, causing various problems in casting and also in final products. Hence, research works were carried out with the attempt to relieve bad effects of high-melting-point inclusions. Usually, solid inclusions in Al deoxidized steel can be modified into low-melting-temperature ones by Ca-treatment. Whereas, large D-type inclusions are not easy to prevent during Ca-treatment. As [Ca] can also be supplied into liquid steel bulk by chemical reactions between steel and basic slag, hence, present authors wanted to find out how to target low-melting-point inclusions in Al deoxidized special steel by slag refining. Before the conduction of a series of laboratorial experiments, it would be desirable to carry out thermodynamic calculations for deeper insights on this topic. Hence, the work in this chapter was done. It was found that targeting low-melting-point inclusions in Al deoxidized steel by slag-steel reaction is feasible, in which the high basicity high alumina refining slag should be used.

In previous Chap. 2, formation of spinel inclusions in high strength low alloy steel were investigated by laboratorial slag-steel reaction experiments. It has known that spinel can be very easily produced during the basic slag refining when certain amount of [Al] was contained in steel. As spinel inclusions were featured by high melting point and poor deformability, they often worsen the castability of steel and deteriorate performances of final products. In previous works, fundamental studies on the formation of spinel were carried out for liquid Fe melt and in stainless steel melts. It was known that spinel inclusions can be produced even [Al] and [Mg] in the metal were of only several parts per million. Moreover, spinel can be modified into calcium aluminates spontaneously when some [Ca] was contained in steel. In the Chap. 2, the present authors also found in laboratorial experiments that spinel would be modified from blocky shape into spherical shape when Ca entered into the inclusion, which would be favorable to diminish the negative influences of spinel on steel performances.

Now that low-melting-point inclusions can be helpful to improve mechanical property of steel, it is worthy to study and find out the way to target low melting point

M. Jiang and X. Wang, *Slag-Steel Reaction and Control of Inclusions in Al Deoxidized Special Steel*, Engineering Materials, https://doi.org/10.1007/978-981-19-3463-6_3

Table 3.1 Chemical compositions of steel, mass%

Element	C	Si	Mn	P	S	Cr
Content	0.60	1.5	0.8	0.005	0.0047	0.20

inclusions in Al deoxidized steel. However, this topic has not discussed intensively enough for Al-killed low alloy steels. In the past, discussions on Al deoxidized steel mainly focused on decreasing total oxygen (T[O]) contents in steel, viz., reducing the amounts of inclusions in steel.

Therefore, in this chapter, formation thermodynamics of inclusions in Al deoxidized steel was conducted to elucidate the way of targeting low-melting-point inclusions with improved deformability by the software Factsage.

3.1 Fundamental Data Related to Calculation

3.1.1 Chemical Compositions of Steel

Chemical compositions of the steel used in the calculation were shown in Table 3.1.

3.1.2 Fundamental Thermodynamic Data

Table 3.2 showed interaction coefficients among the solutes in steel, which were used to estimate activity coefficients of related solutes [1, 2]. The Henry activity coefficient related to the steel sample used in the experiment was calculated according to formula (3.1) and the activity was calculated according to formula.

During the calculation, activity coefficient and activity of each solute element in the steel at 1873 K were obtained by Eqs. (3.1) and (3.2). Where, f_i was the Henry activity coefficients of elements, $a_{[i]}$ was the activity of elements, $e^j{}_i$ was the interaction coefficients of element j to i, while $[i]$ and $[j]$ were mass percentages of elements i and j in molten steel, respectively. With the data in Tables 3.1 and 3.2, Henry activity coefficients of relevant elements at 1873 K can be calculated as Table 3.3.

$$lg f_i = \sum_j e_i^j [j] \tag{3.1}$$

$$a_{[i]} = f_i \cdot [i] \tag{3.2}$$

Table 3.2 Interaction coefficients among the solutes in steel e^i_j at 1873 K [1–3]

j / i	Al	C	O	Ca	Mn	Mg	P	S	Si	Cr	Mo
Al	0.045	0.091	– 6.6	−0.047	0.0065	−0.13	0.033	0.03	0.0056	0.012	–
C	0.043	0.14	−0.34	−0.097	−0.012	0.07	0.051	0.046	0.08	−0.024	– 0.0083
O	−3.9	−0.436	−0.2	−1040	−0.021	−190	0.07	−0.133	0.131	−0.0459	0.0035
Si	0.058	0.18	−0.23	−0.067	0.002	–	0.11	0.056	0.11	−0.0003	–
Mn	–	−0.07	−0.083	−0.023	0.0	–	−0.0035	−0.048	0.0	0.0039	–
Mg	−0.12	−0.24	−289	–	–	–	–	−1.38	−0.09	−0.05	–
Ca	−0.072	−0.34	−9000	−0.002	−0.0156	–	−0.097	−336	−0.097	0.02	–
S	0.035	0.11	– 0.27	−269	−0.026	−1.82	0.29	−0.029	0.063	−0.011	0.0027
P	–	0.13	0.13	–	0.0	–	0.062	0.028	0.12	−0.03	–

Table 3.3 Activity coefficient (f_i) calculated at 1873 K

f_C	f_{Mn}	f_{Si}	f_{Al}	f_O	f_{Ca}	f_{Mg}
1.55	0.91	1.88	1.18	0.22	0.00018	0.46

3.2 Formation Thermodynamics of Inclusions

3.2.1 *Chemical Reaction Equilibrium Between Steel and CaO–SiO₂–Al₂O₃ System*

The phase diagram calculation module of the software was used to calculate the melting points of CaO–Al_2O_3–SiO_2 ternary system to obtain iso-temperature lines. The calculation results were shown in Fig. 3.1. As can be seen, the melting point of region 1′ was lower than 1673 K, which contained about 28%–70% SiO_2, 3%–40% Al_2O_3 and 10%–55% CaO, as shown in the figure by the thick solid line. While the melting point of region 2′ was lower than 1773 K, which contained about 0–15% SiO_2, 22%–55% Al_2O_3 and 42%–65% CaO. In region 2′, if the inclusions contained about 0–10% SiO_2, 35%–50% Al_2O_3 and 47%–57% CaO, melting points of the inclusions would be lower than 1673 K.

In order to calculate the chemical reaction equilibrium of molten steel and CaO–Al_2O_3–SiO_2 system, activity of each component in CaO–Al_2O_3–SiO_2 ternary system at 1873 K should be obtained. The calculated iso-activity curves were plotted in Figs. 3.2, 3.3 and 3.4, respectively.

During the chemical reactions of slag-steel-inclusion, if local equilibrium among them was established, dissolved oxygen ([O]) in steel can be estimated by following reactions and equations, based on the [Si]–[O] in liquid steel [1].

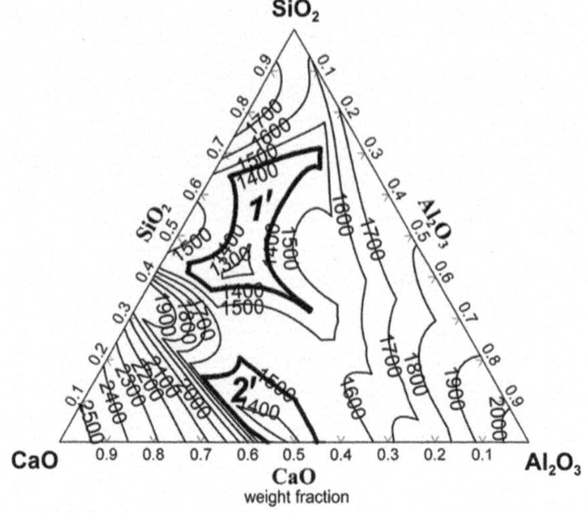

Fig. 3.1 Iso-temperature lines in CaO–Al_2O_3–SiO_2 system

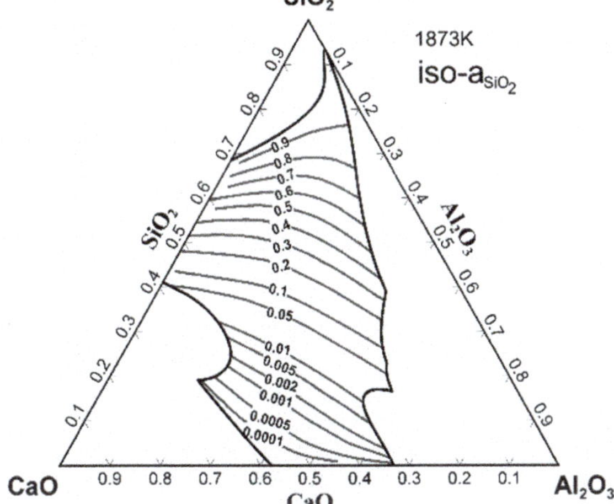

Fig. 3.2 Iso-activity lines of SiO_2 at 1873 K

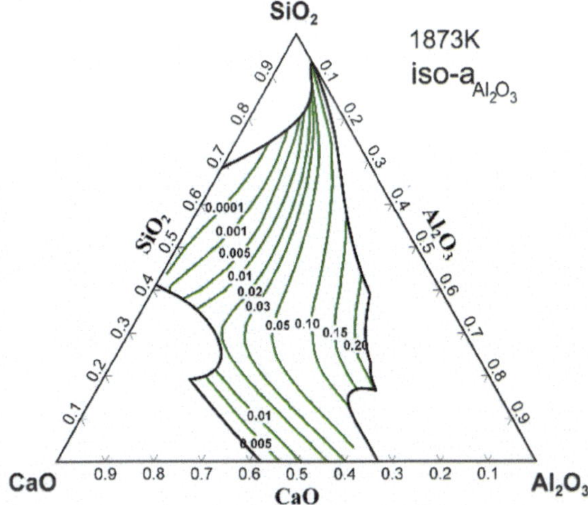

Fig. 3.3 Iso-activity lines of Al_2O_3 at 1873 K

$$[\mathrm{Si}] + 2[\mathrm{O}] = (\mathrm{SiO_2}) \tag{3.3}$$

$$\Delta G^{\Theta} = -581{,}900 + 221.8T\,\mathrm{J/mol} \tag{3.4}$$

Based on Eqs. (3.1), (3.2) and (3.4), (3.5) can be deduced to calculate the content of [O]. Where, activities of [Si] and [O] can be calculated based on the compositions of steel while the activity SiO_2 in slag can be calculated by the FactSage software.

$$[\%\mathrm{O}] = \left(\frac{a_{\mathrm{SiO_2}}}{K \times f_{\mathrm{Si}} \times [\mathrm{Si}] \times f_{\mathrm{O}}^2}\right)^{1/2} \tag{3.5}$$

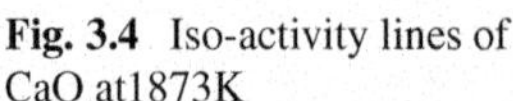

Fig. 3.4 Iso-activity lines of CaO at1873K

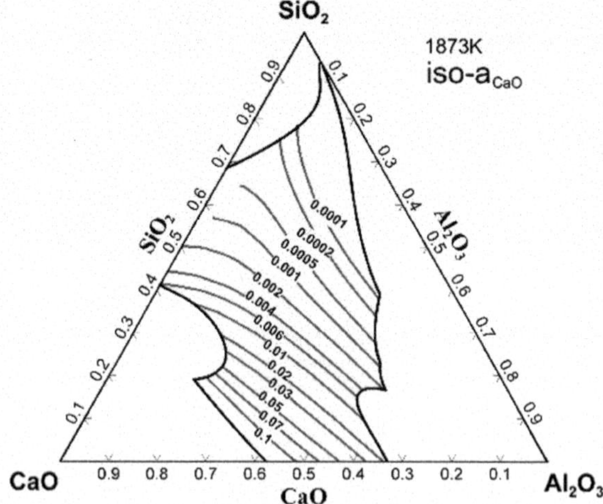

Based on the activity data of SiO_2 in Fig. 3.2 and fundamental data in Tables 3.1 and 3.3, iso-[O] lines can be plotted as Fig. 3.5. As a result, at 1873 K, changes of [O] contents in steel equilibrated with the slag or inclusions composed of CaO–SiO_2–Al_2O_3 system can be clearly seen.

It can be seen from Fig. 3.5 that the dissolved oxygen contents in liquid steel equilibrated with low melting point region 1′ was about 0.0040%–0.012%, while the [O] contents in liquid steel equilibrated with region 2′ was much lower and in the range of 0.0001%–0.0003%. With the increase of slag basicity CaO/SiO_2, [O] contents in molten steel decreased obviously. With mass percentage ratio of CaO/Al_2O_3 kept constant, the content of [O] in molten steel decreased obviously with the decrease of SiO_2 content in slag, too.

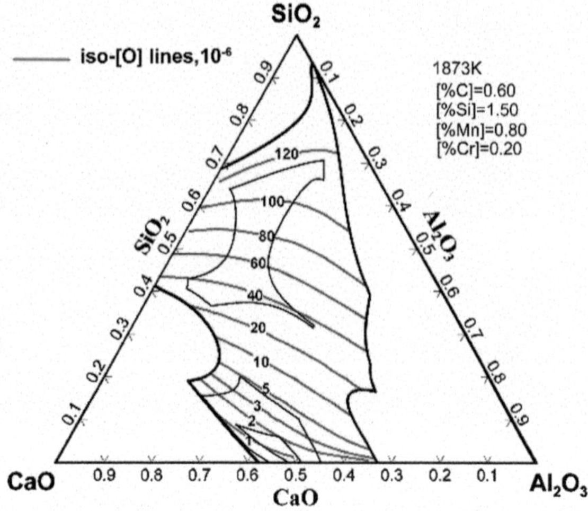

Fig. 3.5 Iso-[O] lines in steel equilibrated with CaO–SiO_2–Al_2O_3 system

Equilibration of [Al] in steel with $CaO–SiO_2–Al_2O_3$ slag/inclusions can be expressed as following [4]:

$$4[\mathrm{Al}] + 3(\mathrm{SiO_2}) = 2(\mathrm{Al_2O})_3 + 3[\mathrm{Si}] \tag{3.6}$$

$$\Delta G^{\Theta} = -658{,}300 + 107.2T\,\mathrm{J/mol} \tag{3.7}$$

Through Eqs. (3.1), (3.2) and (3.7), formula to calculate the contents of [Al] in steel can be deduced as following, when liquid steel equilibrated with the $CaO–SiO_2–Al_2O_3$ system.

$$[\%\mathrm{Al}] = \left(\frac{a^2_{\mathrm{Al_2O_3}} \times f^3_{[\mathrm{Si}]} \times [\%\mathrm{Si}]^3}{a^3_{\mathrm{SiO_2}} \times f^4_{[\mathrm{Al}]} \times K}\right)^{1/4} \tag{3.8}$$

Substituting the activities of SiO_2 and Al_2O_3 of $CaO–SiO_2–Al_2O_3$ ternary system in Figs. 3.2 and 3.3, as well as the relevant data in Tables 3.1 and 3.2 into Eq. (3.8), the iso-[Al] lines in molten steel can be plotted as Fig. 3.6.

It can be seen from Fig. 3.6 that the [Al] contents in liquid steel equilibrating with low melting point region 1′ were in the range of 0.0001–0.0010%, while the [Al] contents in liquid steel equilibrating with lower melting point region 2′ were much higher and in the scope of 0.0300–0.10%. Moreover, the equilibrium [Al] content in steel increased with the rise of CaO/SiO_2 ratio and CaO/Al_2O_3 ratio.

Similarly, contents of [Ca] in liquid steel equilibrated with $CaO–SiO_2–Al_2O_3$ system can be calculated from the following reaction and equations.

$$2[\mathrm{Ca}] + (\mathrm{SiO_2}) = 2(\mathrm{CaO}) + [\mathrm{Si}] \tag{3.9}$$

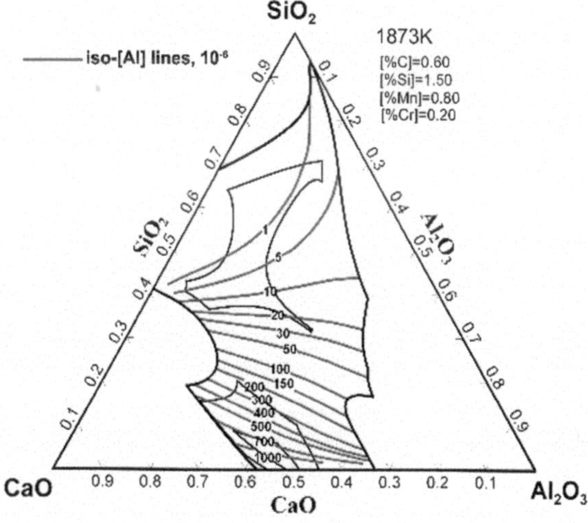

Fig. 3.6 Iso-[Al] lines in steel equilibrating with $CaO–SiO_2–Al_2O_3$ system

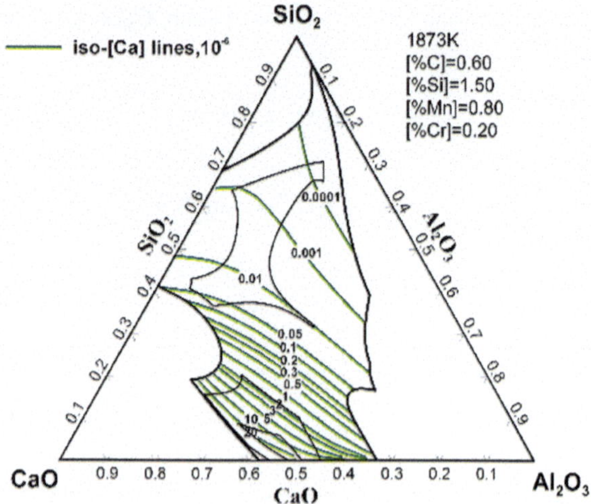

Fig. 3.7 Iso-[Ca] lines in steel equilibrating with $CaO–SiO_2–Al_2O_3$ system

$$\Delta G^{\Theta} = -694{,}422 + 75.06T\,\mathrm{J/mol} \tag{3.10}$$

Through Eqs. (3.1), (3.2)and (3.10), the formula to calculate [Ca] in steel can be deduced as:

$$a_{[\mathrm{Ca}]} = \left(\frac{a_{\mathrm{CaO}}^2 \times f_{\mathrm{Si}} \times [\%\mathrm{Si}]}{a_{\mathrm{SiO_2}} \times K} \right)^{1/2} \tag{3.11}$$

With the calculated activities of CaO and SiO_2 in the ternary system of $CaO–SiO_2–Al_2O_3$ in Figs. 3.4 and 3.2 and relevant data in Tables 3.1and 3.2, the iso-[Ca] lines in liquid steel of equilibrated with $CaO–SiO_2–Al_2O_3$ ternary system at 1873 K can be drawn as Fig. 3.7.

It can be seen from Fig. 3.7 that [Ca] contents in liquid steel equilibrating with low-melting-point region 1′ was very low and was less than 0.01 ppm, while [Ca] content in liquid steel equilibrating with low melting point region 2′ was relatively higher, ranging from 0.0003 to 0.0015%. Keeping the ratio of CaO/Al_2O_3 constant, [Ca] contents in steel would increase with the rise of CaO/SiO_2 ratio.

When local equilibrium of slag-steel-inclusion was established, chemical reaction equilibrium of [Al] and [Si] in steel can be established through the reaction shown in (3.6). Therefore, the relation of [Al] and on [Si] content in liquid steel can be obtained. When [Al] contents were 0.04%, 0.05%, 0.06% and 0.07%, iso-[Si] isolines in steel equilibrated with slag or inclusion with chemistry locating in the lower melting point region 2′ were calculated and plotted as Fig. 3.8.

It can be seen from Fig. 3.8, with the rise of [Al] contents in steel, [Si] contents in steel increased from 0.0400% to 0.0007%. At a constant content of [Al], concentrations of [Si] would increase with the rise of SiO_2 contents in slag. With the contents

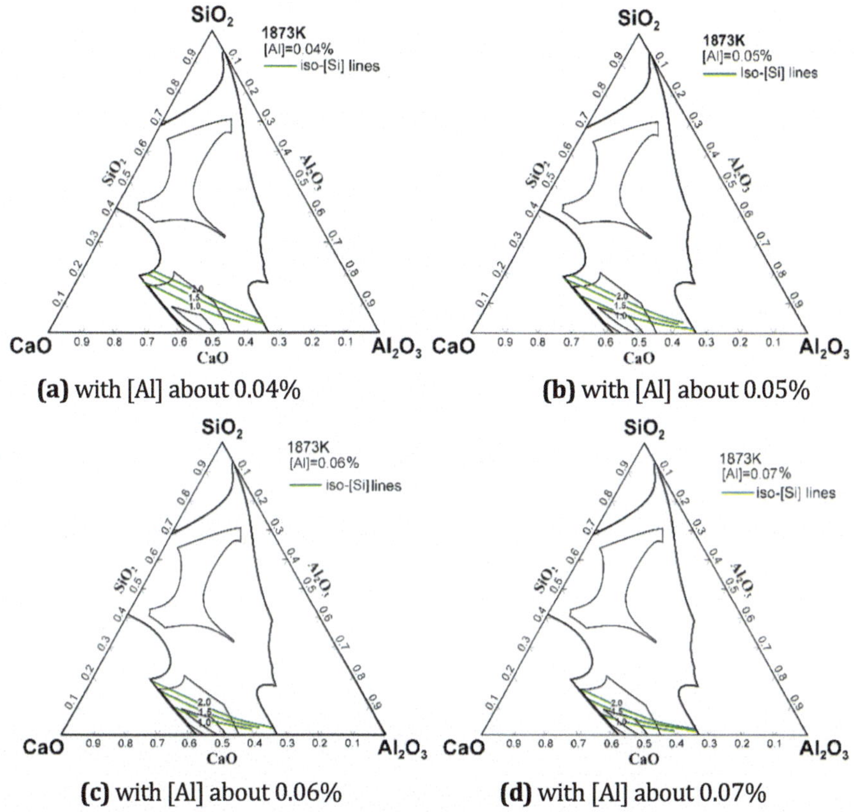

(a) with [Al] about 0.04%

(b) with [Al] about 0.05%

(c) with [Al] about 0.06%

(d) with [Al] about 0.07%

Fig. 3.8 Iso-[Si] lines equilibrating with varied [Al] in steel

of [Al] in steel about 0.0500%, contents of [Si] in steel equilibrating with the lower melting point region 2′ can be kept in the range of 1.5–2.0%.

3.2.2 Chemical Reaction Equilibrium of Steel and CaO–SiO$_2$–Al$_2$O$_3$–MgO System

Nowadays, MgO-based refractories are widely used as linings of metallurgical vessels and reactors. On one hand, at the temperature of steelmaking, slag line of the ladle would be eroded by refining slag. As a result, refining slags in steelmaking process usually contain certain amounts of MgO (often about 10% or so). Hence, refining slags are composed of CaO–SiO$_2$–Al$_2$O$_3$–MgO quaternary system. Considering that, chemical reaction equilibrium of steel and CaO–SiO$_2$–Al$_2$O$_3$–10%MgO system was also evaluated in this section. First, melting temperature of this quaternary was calculated to obtain the information of low melting point regions, as shown

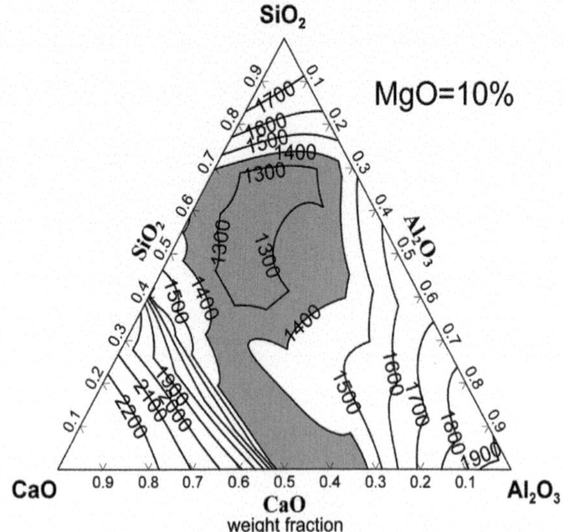

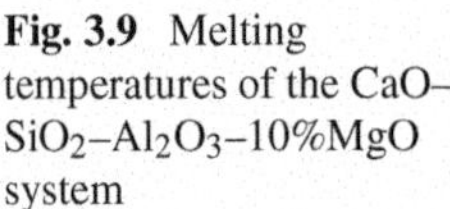
Fig. 3.9 Melting temperatures of the CaO–SiO_2–Al_2O_3–10%MgO system

in Fig. 3.9, in which the shadowed regions referred to the low melting point zones (<1673 K). As can be seen, contents of MgO significantly influenced the melting temperature of this system.

The calculated activities of components (CaO, SiO_2, Al_2O_3 and MgO) in this quaternary system were indicated by the iso-activity lines in the following Fig. 3.10.

Utilizing the calculated activity data in Fig. 3.10 and thermodynamic data in Tables 3.2 and 3.3, contents of [O], [Al], [Ca] and [Mg] in steel equilibrating with CaO–SiO_2–Al_2O_3–10%MgO system can be obtained to draw the iso-[O], iso-[Al] and iso-[Ca] lines, as shown in Fig. 3.11.

3.2.3 Formation of Low Melting Point Calcium Magnesia Aluminate Inclusions

According to previous studies, it had been known that spinel inclusions can be easily formed when [Mg] was contained in liquid steel, even if the mass percentages of [Al] and [Mg] were as low as several parts per million. Moreover, when the steel contained certain amounts of [Ca], spinel inclusions would be transformed into calcium magnesia aluminate together with limited content of SiO_2.

According to calculation conducted above, it can be known that, to target inclusion in the lower melting point region 2′ in the CaO–SiO_2–Al_2O_3 ternary system, dissolved oxygen content in steel should be controlled in the range of 0.0001–0.0003%, while the content of [Ca] should be controlled in the scope of 0.0003–0.0015%. Also, the content of [Al] should be in the range of 0.04–0.06%, which meant that strong Al deoxidation should be used in steelmaking process. To ensure accurate control of

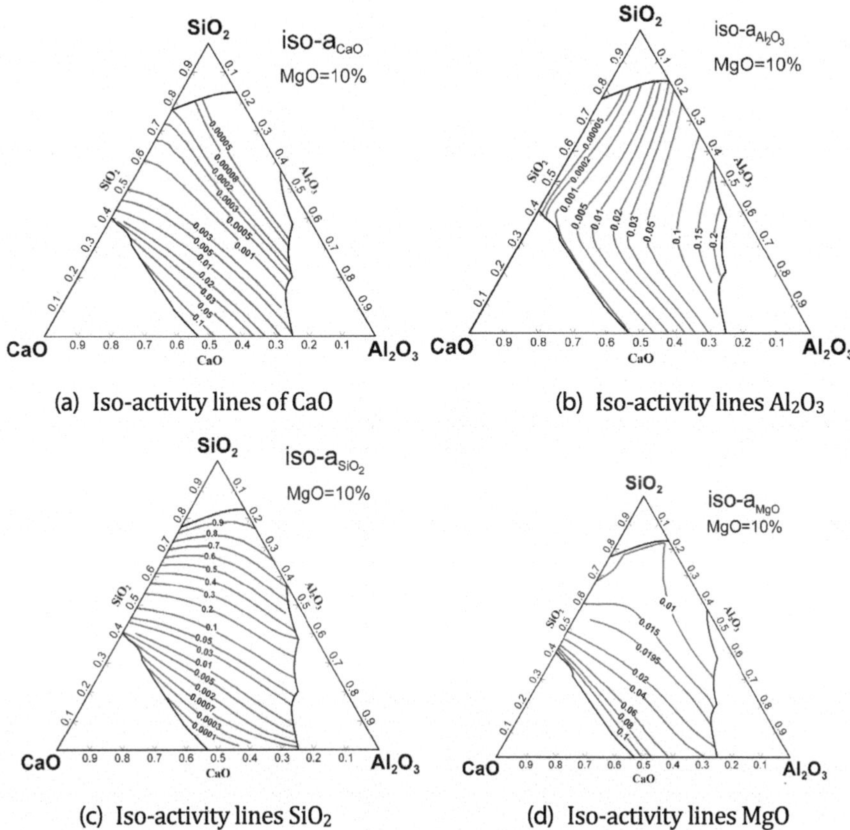

(a) Iso-activity lines of CaO

(b) Iso-activity lines Al_2O_3

(c) Iso-activity lines SiO_2

(d) Iso-activity lines MgO

Fig. 3.10 Iso-activity lines in $CaO–SiO_2–Al_2O_3–10\%MgO$ system

[Si] in steel, equilibrium between [Al] and [Si] was also evaluated, which proved that the [Si] content in liquid steel can be well controlled within the targeted range. To achieve that, the quaternary system of $CaO–SiO_2–Al_2O_3–10\%MgO$ equilibrating with liquid steel should be featured by high basicity and high alumina contents.

Therefore, [Al], [Ca] and [O] equilibrating with $CaO–SiO_2–Al_2O_3–10\%MgO$ slag system was evaluated, as shown in Figs. 3.12, 3.13 and 3.14. It can be seen that, to target the contents of [Al] in the range of 0.04%–0.06%, contents of Al_2O_3 and CaO in slag should be in the range of 35%–48% and 36%–45%, respectively. To target the contents of [O] in the range of 0.0001–0.0003%, contents of Al_2O_3 and CaO in slag should be about 35%–45% and 38%–48%, respectively. To target the content of [Ca] in the range of 0.0003%–0015%, contents of Al_2O_3 and CaO in slag should be about 38%–45% and 39%–45%, respectively. To sum, high basicity high alumina slag system should be used to target low melting point inclusion in Al deoxidized low alloy steel.

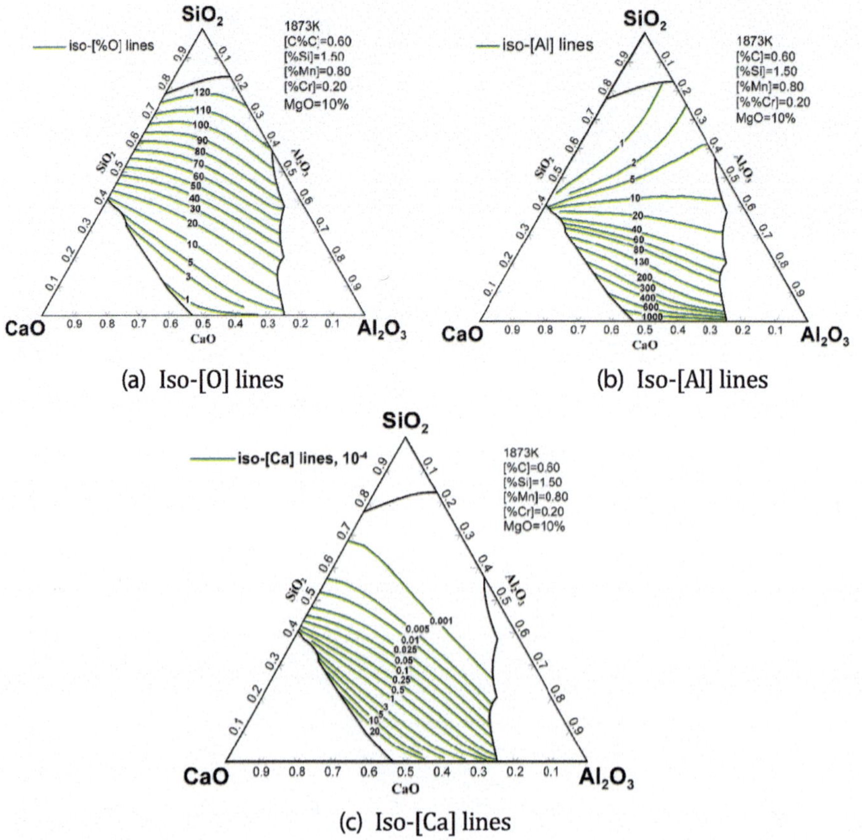

(a) Iso-[O] lines

(b) Iso-[Al] lines

(c) Iso-[Ca] lines

Fig. 3.11 Iso-concentration lines of [Al], [Ca] and [O] equilibrating with $CaO–SiO_2–Al_2O_3$–10%MgO system

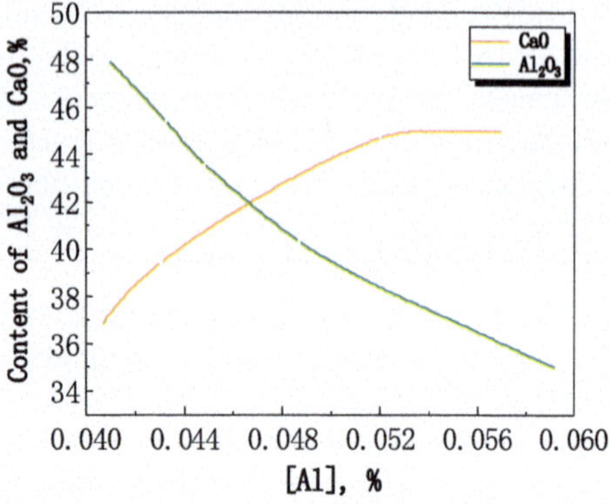

Fig. 3.12 Influences of CaO and Al_2O_3 contents in Slag on [Al] in steel

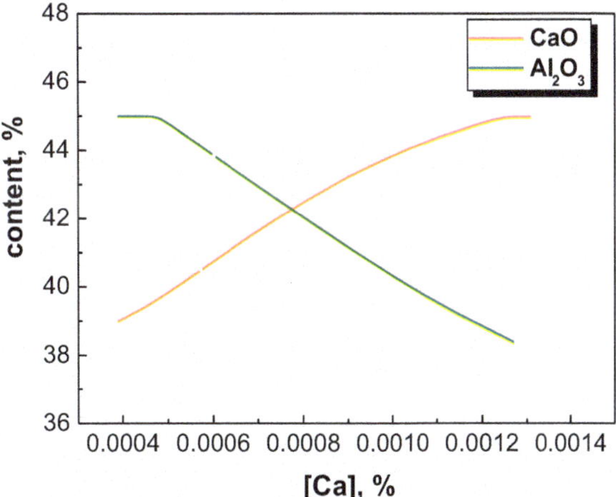

Fig. 3.13 Influences of CaO and Al_2O_3 contents in Slag on [Ca] in steel

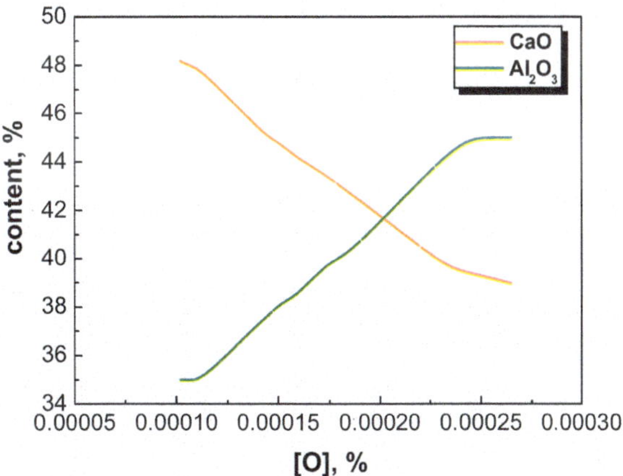

Fig. 3.14 Influences of CaO and Al_2O_3 contents in Slag on [O] in steel

3.3 Summary

By calculating the formation thermodynamics of inclusions based on chemical reaction equilibrium among slag-steel-inclusions, following conclusions can be obtained:

(1) In order to improve the fatigue property of steel and reduce the oxygen content of liquid steel, it is necessary to target the inclusions in the lower melting point region 2′ in the $CaO–Al_2O_3–SiO_2$ system, with SiO_2, Al_2O_3 and CaO about 0–15%, 32%–55% and 42%–61%, respectively. To achieve that, [Al] contents in the liquid steel should be in the range of 0.035%–0.10% while the content of [Ca] should be in the scope of 0.0003%–0.0015%and the dissolved oxygen content is needed to controlled in the range of 0.0001%–0.0003%.

(2) The equilibrium relationship between [Al] and [Si] would be established when local chemical reaction equilibrium among slag-steel-inclusion was approached. With the rise of [Al] contents from 0.040% to 0.070%, [Si] contents in steel increased. To target proper [Si] contents in spring steel, it is necessary to accurately control the contents of [Al].
(3) The thermodynamic calculation showed that, by appropriate selection of refining slag system, inclusion in steel can be controlled in the lower melting point region 2′ and simultaneously with proper control of [Si] content in steel.
(4) In order to target low melting point inclusions in Al deoxidized low alloy steel, high basicity high alumina content refining slag system should be used.

References

1. Sigworth, G.K., Elliott, J.F.: The thermodynamics of liquid dilute iron alloys. Metal Science **8**, 298–310 (1974)
2. Chen, X.W.: Deoxidation in Steelmaking, 11. Metallurgy Industry Press, Beijing (1991)
3. Suito, H., Inoue, R.: Thermodynamics on control of inclusions composition in ultraclean steels. ISIJ Int. **36**(5), 528–536 (1996)
4. Ohta, H., Suito, H.: Activities in CaO-SiO_2-Al_2O_3 slags and deoxidation equilibria of Si and Al. Metall. and Mater. Trans. B. **27B**(4), 943–953 (1996)

Chapter 4
Slag-Steel Reaction and Control of Low-Melting-Point Inclusions in Al-Deoxidized Steel

Abstract In this chapter, formation of low-melting-point inclusions in Al-deoxidized steel melts was discussed by a series of slag-steel reaction experiments in laboratory. Two types of high basicity high alumina slags, with initial basicity about 8 while alumina contents about 45% and 35%, respectively, were used in the experiments. It was found that high cleanliness can be achieved in steel melts, with T.O contents about 0.0010% or lower. Besides, low-melting-point inclusions were successfully targeted in steel. Based on the obtained results, influences of alumina content in the used slags on the control of inclusions and the relationship between the compositions of slag and the formed inclusions were also discussed.

According to the formation thermodynamic calculation of inclusions in previous chapter, it can be known that under the condition of Al deoxidation, in order to target calcium magnesia aluminate with low-melting-point in the spring steel, the refining slag system would be featured by high basicity and high content of Al_2O_3.

Based on thermodynamic calculation, the control of low melting point inclusions in Al deoxidized spring steel was studied and verified by a series of designed laboratorial slag-steel chemical reaction experiments, during which the used refining slags were with high basicity of 8 while the alumina content was about 45% and 35%, respectively.

4.1 Slag-Steel Reaction Experiments in Lab

4.1.1 Preparation of Steel and Slag Materials

In order to carry out the laboratorial slag-steel reaction experiments, master metal was firstly prepared in a vacuum induction furnace, which were with [C], [Si], [Mn], [Cr], [P] and [S] in the range of 0.56%–0.64%, 1.50%–2.0%, 0.60%–0.90%, 0.08%–0.15%, < 0.035% and < 0.035%, respectively. Two kinds of master metals with different Al contents were prepared, hence, initial aluminum contents in steel can

M. Jiang and X. Wang, *Slag-Steel Reaction and Control of Inclusions in Al Deoxidized Special Steel*, Engineering Materials, https://doi.org/10.1007/978-981-19-3463-6_4

Table 4.1 Chemical compositions of master metal (mass%)

	Elements					
	C	Si	Mn	S	Cr	Al
Low Al master metal	0.59	1.53	0.82	0.0047	0.18	0.0050
High Al master metal	0.58	1.90	0.80	0.0047	0.19	0.0750

be adjusted by proper mixing of the high-aluminum and the low-aluminum master metals. Chemical compositions of the master metals were shown in Table 4.1.

Slag used in the experiments was composed of CaO–Al_2O_3–SiO_2. The used raw materials were high purity (purity is 99.99%) CaO, Al_2O_3 and SiO_2 of chemical analysis grade. After fully mixing with each other, the prepared slag would be used in the slag-steel reaction experiments.

4.1.2 Experimental Procedures

An electrical resistance vertical furnace was used in the slag-steel reaction experiments. The furnace can be heated to about 2073 K and the structure of it was schematically given in Fig. 4.1. The R-type thermocouple was used as to control the temperature. During the slag-steel reaction experiments, master metal and slag were put together into the MgO crucible. The crucible was then put into the corundum reaction tube and set at the even temperature zone of the furnace. Afterwards, high purity argon gas was introduced into the tightly sealed reaction tube for evacuation about 5 min. Afterwards, the heating would be started to raise the temperature to about 1873 K $\pm$ 2 K and the melts would be held at this temperature for determined period of time. At the end of the holding, the crucible would be quickly taken out of the furnace and put into water for quenching.

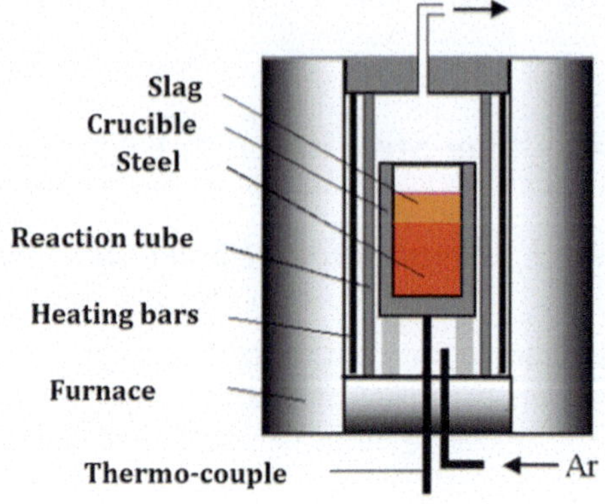

Fig. 4.1 Schematics of the electrical resistance furnace

To elucidate the influences [Al] contents in steel, basicity and initial Al_2O_3 contents of refining slag on the formation of calcium magnesium aluminate inclusions, the designed laboratorial experiments were mainly composed of two parts. In the first part of experiments, the influence of slag-steel reaction on the formation of inclusions were explored for the slag with basicity about 8 while alumina content about 45% (named slag A for conveniece of discussion). In the other part of experiments, the influence of slag with basicity about 8 while alumina about 35% (name slag B for conveniece of discussion) on the formation of inclusions were studied (Table 4.2).

Table 4.2 Experimetal scheme of the slag-steel experiments in lab

	Heat	High-Al master metal, g	Low-Al master metal, g	Slag	Initial [Al], mass%
Slag A	1	100	0	CaO/SiO_2: 8; Al_2O_3: 45%	0.0750
	2	92.9	7.1		0.0700
	3	85.8	14.2		0.0650
	4	78.6	21.4		0.0600
	5	71.5	28.5		0.0550
	6	64.4	35.6		0.0500
	7	57.3	42.7		0.0450
	8	50.1	49.9		0.0400
	9	43.0	57.0		0.0350
	10	35.9	64.1		0.0300
	11	28.8	71.2		0.0250
	12	21.7	78.3		0.0200
	13	14.5	85.5		0.0150
	14	7.4	92.6		0.0100
	15	0	100		0.0050
Slag B	16	0	100	CaO/SiO_2: 8; Al_2O_3: 35%	0.0050
	17	7.4	92.6		0.0100
	18	14.5	85.5		0.0150
	19	21.7	78.3		0.0200
	20	28.8	71.2		0.0250
	21	35.9	64.1		0.0300
	22	43.0	57.0		0.0350
	23	50.1	49.9		0.0400
	24	57.3	42.7		0.0450
	25	64.4	35.6		0.0500
	26	71.5	28.5		0.0550
	27	100	0		0.0750

4.2 Determination of Slag-Steel Reaction Time

As mentioned above, slag-steel should react with a certain period of time at 1873 K before the melt was taken out of the furnace for quenching. Hence, before the two series of experiments, experiments should be carried out to determine needed time for slag and steel to approach local chemical reaction equilibrium. As a result, the pre-equilibrium experiment was carried out, during which steel about 100 g, slag about 50 g (with basicity about 8 and alumina content about 45%) were charged into MgO crucibles. In the experiment, the initial [Al] content in steel was set as 0.0050%. After the slag and steel reacted for 30 min, 60 min, 90 min, 120 min and 180 min, respectively, MgO crucible were taken out of the furnace and quenched in water. The obtained steel samples and slag samples were then prepared for chemical composition analysis.

The results of chemical composition of steel sample and slag sample were shown in Tables 4.3 and 4.4.

According to Tables 4.3 and 4.4, Figs. 4.2 and 4.3 can be obtained. It can be seen that after 90 min slag-steel reaction, contents of [Al] and T[O] in molten steel and the contents of Al_2O_3 in slag kept very constantly with the increase of reaction time. Hence, it was considered that chemical reaction of slag-steel at 1873 K can approach local equilibrium within 90 min.

Table 4.3 Chemical compositions of steel samples in pre-equilibrium experiments (mass%)

Reaction time (min)	C	Si	Mn	S	Cr	Al	T[O]	N
30	0.46	1.52	0.84	0.0005	0.17	0.038	0.0014	0.0017
60	0.37	1.52	0.84	0.0006	0.17	0.041	0.0012	0.0018
90	0.46	1.50	0.84	0.0008	0.17	0.042	0.0008	0.0015
120	0.36	1.48	0.84	0.0007	0.17	0.043	0.0007	0.0018
180	0.46	1.52	0.84	0.0005	0.17	0.040	0.0007	0.0017

Table 4.4 Chemical compositions of slag samples in pre-equilibrium experiments (mass%)

Reaction time (min)	CaO	Al_2O_3	SiO_2	MgO	MnO	FeO
30	44.90	40.50	6.90	6.56	0.045	0.72
60	44.30	40.13	6.26	8.61	0.038	0.26
90	42.68	39.37	6.95	9.67	0.042	0.24
120	42.95	39.74	6.23	10.62	0.048	0.28
180	42.43	39.47	6.57	10.90	0.048	0.48

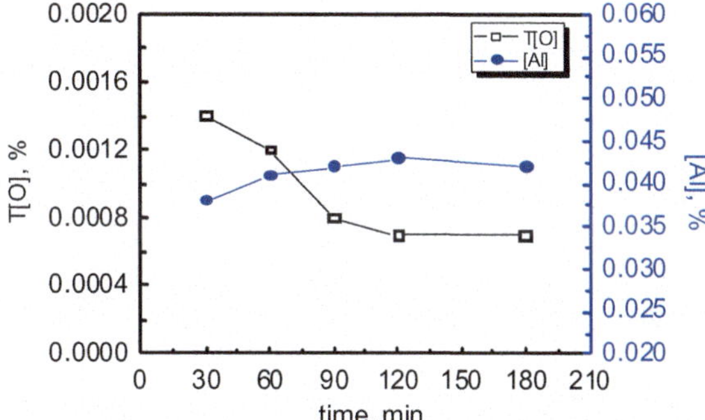

Fig. 4.2 Variations in the contents of [Al] and T.O in steel melts against reaction time

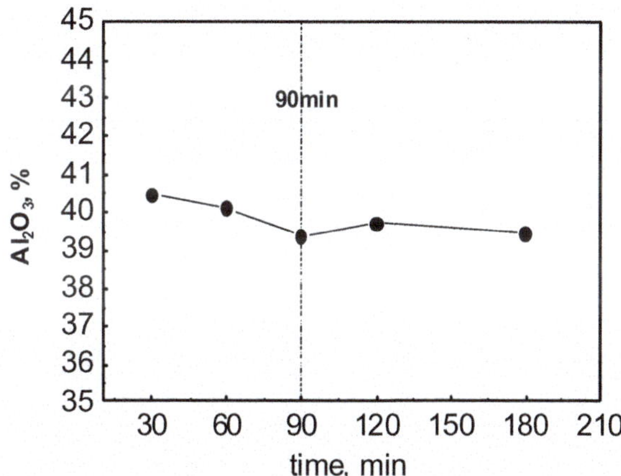

Fig. 4.3 Variations in the contents of Al_2O_3 in slag with reaction time

4.3 Control of Inclusions in Steel Refined by High Basicity High Alumina Slag A

After determination of reaction time (90 min), slag-steel reaction experiments were carried out. Slag A with basicity about 8 while alumina content about 45% was used and 15 heats of experiments were carried out. During the experiments, the effect of initial [Al] contents in steel on the control of inclusions in steel was taken into account, in which initial [Al] content varied in the range of 0.0050–0.0750%. After each experiment, chemical composition of the steel and slag samples were analyzed and inclusions in steel samples were observed carefully under the SEM–EDS machine.

4.3.1 Chemical Compositions of Steel and Slag

Contents of [C], [Si], [Mn], [S] and [Cr] in steel were in the scope of 0.45%–0.49%, 1.36%–1.94%, 0.79%–0.87%, 0.0005%–00.007% and 0.17%–0.20%, respectively. While the contents of strong deoxidation elements of [Al], [Ca] and [Mg] were in the range of 0.030%–0.047%, 0.0005%–0.0012% and 0.0010%–0.0014%, respectively. Moreover, steel melts were featured by high cleanliness, with [N] and T[O] about 0.0009%–0.0020% and 0.0007%–0.0012%, respectively, as shown in Table 4.5.

After 90 min of reaction, contents of CaO, SiO_2, Al_2O_3, MgO, FeO and MnO in slag were in the ranges of 38.71%–42.78%, 5.61%–6.14%, 37.13%–41.7%, 9.60%–10.88%, 0.14%–0.40% and 0.039%–0.10%, respectively, as shown in Table 4.6.

Distribution of lag compositions in the (CaO + MgO)–SiO_2–Al_2O_3 pseudo-ternary system can be given as Fig. 4.4. It can be seen that slag compositions were in the lower melting point region composed of neighbor to $3CaO{\cdot}Al_2O_3$, $CaO{\cdot}Al_2O_3$ and $12CaO{\cdot}7Al_2O_3$. The mass percentage ratios of (CaO + MgO)/SiO_2 in slag were mainly in the range of 8.5–9.5 while Al_2O_3 contents of in slag were in the range of 40%–50%, as shown in Fig. 4.5.

When chemical reaction equilibrium of molten steel and slag was approached, MgO contents in slag was kept about 10% with very limited fluctuations, as shown in Fig. 4.6. It can be inferred that dissolution of MgO from crucible into slag has reached saturation. With the increase of (CaO + MgO)/SiO_2 ratios in slag, contents of (FeO + MnO) in slag sharply decreased. When the ratios of (CaO + MgO)/SiO_2 reached 9.0–9.5, contents of (FeO + MnO) in slag decreased to 0.1%–0.2%. In the

Table 4.5 Chemical compositions of steel samples refined by slag A

Heat	C	Si	Mn	Cr	S	Al	Ca	Mg	T[O]	N
1	0.46	1.36	0.80	0.17	0.0007	0.030	0.0006	0.0011	0.0007	0.0014
2	0.45	1.46	0.79	0.19	0.0005	0.042	0.0006	0.0012	0.0009	0.0015
3	0.47	1.44	0.87	0.20	0.0006	0.036	0.0009	0.0012	0.0010	0.0010
4	0.47	1.40	0.84	0.19	0.0006	0.038	0.0005	0.0012	0.0012	0.0012
5	0.47	1.49	0.81	0.19	0.0007	0.040	0.0009	0.0012	0.0007	0.0020
6	0.47	1.46	0.80	0.19	0.0005	0.042	0.0012	0.0014	0.0009	0.0012
7	0.46	1.66	0.84	0.18	0.0006	0.038	0.0007	0.0014	0.0009	0.0014
8	0.49	1.68	0.84	0.18	0.0006	0.036	0.0006	0.0010	0.0007	0.0015
9	0.48	1.76	0.86	0.18	0.0006	0.038	0.0008	0.0010	0.0008	0.0015
10	0.48	1.80	0.83	0.18	0.0006	0.040	0.0006	0.0011	0.0008	0.0015
11	0.46	1.78	0.83	0.19	0.0006	0.038	0.0008	0.0011	0.0008	0.0015
12	0.45	1.82	0.82	0.19	0.0006	0.044	0.0006	0.0012	0.0007	0.0013
13	0.48	1.89	0.80	0.19	0.0006	0.044	0.0008	0.0012	0.0009	0.0012
14	0.48	1.94	0.82	0.19	0.0006	0.047	0.0008	0.0014	0.0009	0.0010
15	0.49	1.78	0.79	0.18	0.0007	0.046	0.0005	0.0013	0.0010	0.0009

Table 4.6 Chemical compositions of slag samples

Heat	Components							
	MgO	Al_2O_3	SiO_2	CaO	MnO	FeO	(%S)	MnO + FeO
1	10.88	41.33	6.16	41.92	0.04	0.4	0.016	0.44
2	10.10	40.55	5.72	42.61	0.044	0.14	0.017	0.184
3	10.20	40.32	5.98	42.14	0.10	0.34	0.016	0.44
4	10.16	40.89	6.14	41.68	0.086	0.40	0.015	0.486
5	9.68	40.87	5.74	42.78	0.056	0.25	0.015	0.306
6	18.46	37.13	5.30	38.71	0.049	0.23	0.015	0.279
7	10.82	40.39	5.86	42.09	0.054	0.26	0.016	0.314
8	9.71	40.08	5.81	42.22	0.051	0.23	0.017	0.281
9	9.95	40.41	5.84	42.33	0.049	0.25	0.017	0.299
10	9.68	40.71	5.76	42.70	0.040	0.19	0.016	0.23
11	9.96	41.00	5.93	42.30	0.046	0.22	0.016	0.266
12	10.21	41.70	5.84	42.52	0.041	0.20	0.017	0.241
13	9.60	40.26	5.61	42.60	0.039	0.21	0.017	0.249
14	9.79	41.04	5.78	42.60	0.040	0.20	0.017	0.24
15	10.68	40.44	6.03	42.87	0.041	0.35	0.016	0.391

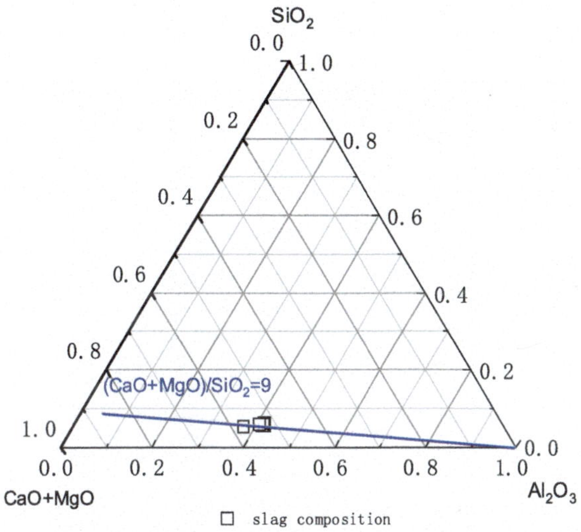

Fig. 4.4 Chemical compositions of slag samples after the experiments (with the use of slag A)

experiments, except for one heat, contents of (FeO + MnO) in slag was always in the range of 0.1–0.3%, as shown in Fig. 4.7.

Because of equilibrium between slag and steel, contents of [Al] in steel were controlled in the range of 0.030–0.047% at the end of experiments, although the initial contents of [Al] widely varied in the range of 0.005%–0.075%. Figures 4.8

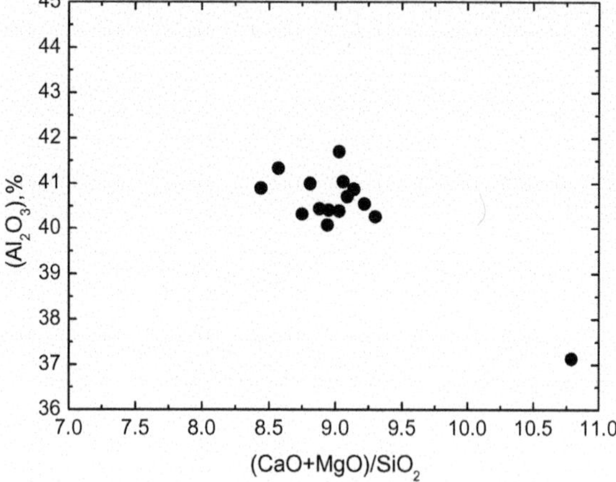

Fig. 4.5 Ratios of $(CaO + MgO)/SiO_2$ and contents of Al_2O_3 in slag

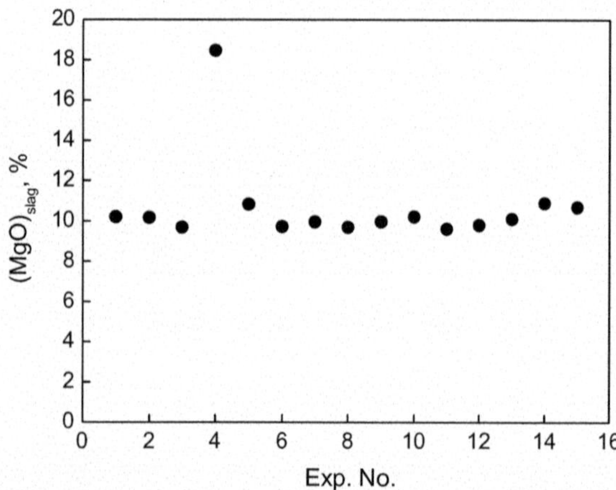

Fig. 4.6 Contents of MgO in slag A

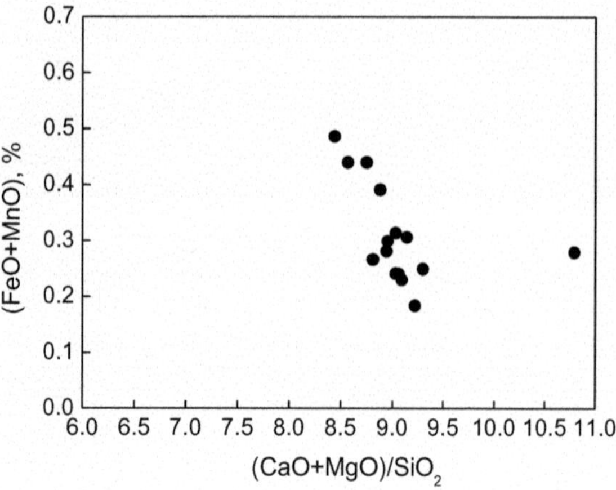

Fig. 4.7 Ration of $(FeO + MnO)$ and $(CaO + MgO)/SiO_2$

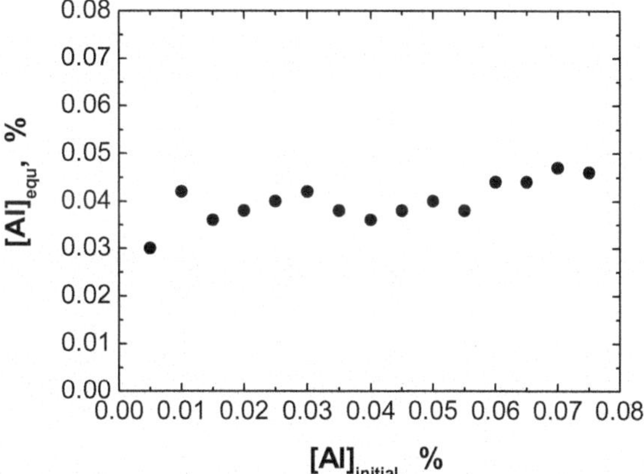

Fig. 4.8 Relation between initial [Al] and final [Al] contents

and 4.9 showed the relationship between equilibrium [Al] contents and initial [Al] contents in different heats. It can be seen that though the initial [Al] contents varied in a wide range, [Al] contents at the end of experiments always fluctuated around a value of 0.0400%. Hence, it can be concluded that the equilibrium [Al] content was slightly affected by the initial [Al] content. As the calculated [Al] content of molten steel equilibrating with the slag was in the range of 0.040–0.060%, it can be known that the experimental results were consistent to thermodynamic calculations, as shown in Fig. 4.10.

As given above in Table 4.4, T[O] contents in steel fluctuated in the range of 0.0006–0.0012% while mainly concentrated in the scope of 0.0007–0.0010%, indicating very high cleanness. As shown in Fig. 4.11, with the rise of $(CaO + MgO)/SiO_2$ ratios from 8.5 to 9.5, T[O] showed a continuous declining tendency. As it known, contents of (FeO + MnO) in slag affected the diffusion deoxidation of molten steel, which in turn affected the cleanness of steel. In the experiments, the content of (FeO

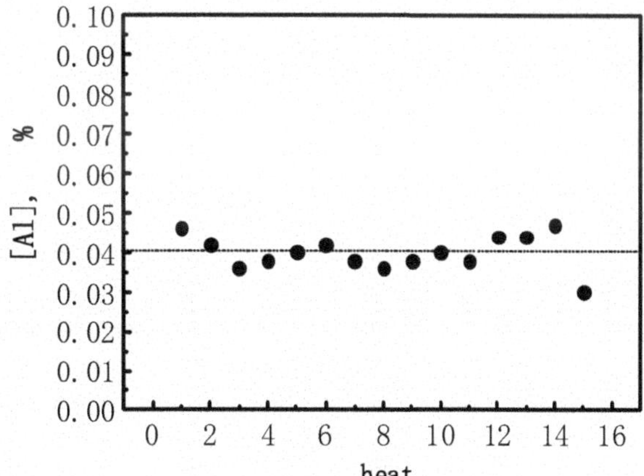

Fig. 4.9 Equilibrium contents of [Al] of all the heats of experiments

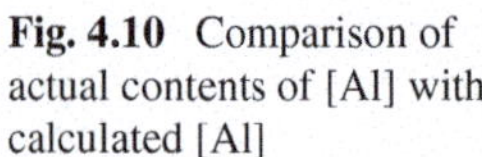

Fig. 4.10 Comparison of actual contents of [Al] with calculated [Al]

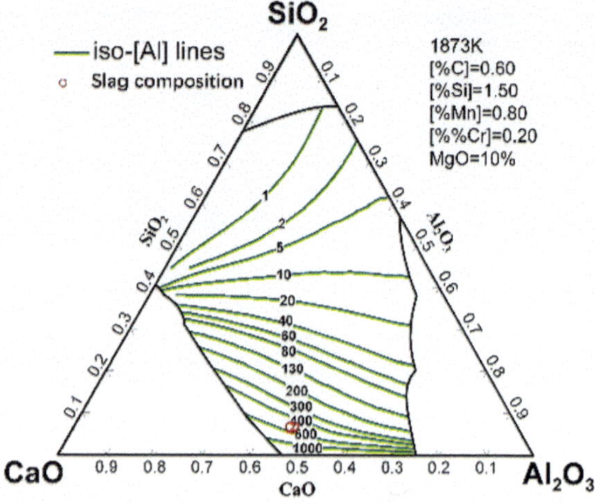

Fig. 4.11 Relation of slag basicity and T[O]

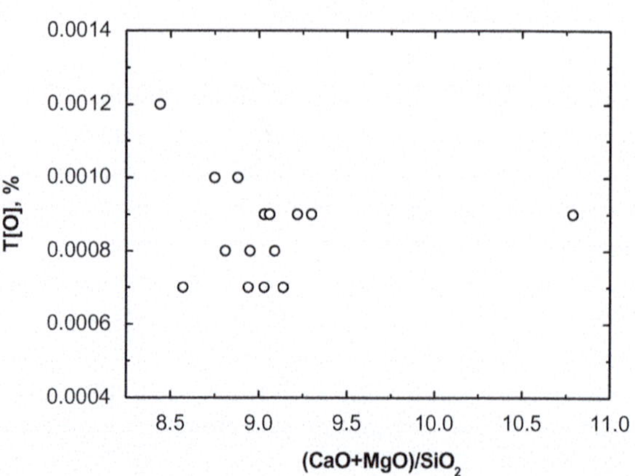

+ MnO) in slag were less than 0.5% and mainly concentrated in the range of 0.2%–0.3%. Correspondingly, the content of T[O] in steel was about 0.0006%–0.0010%, as shown in Fig. 4.12. With the increase of (FeO + MnO) contents from 0.35% to 0.45%–0.50%, T[O] increased from 0.0010% to 0.0010%–0.0012%. It meant that slight fluctuations of (FeO + MnO) contents in slag within 1% can affect T[O] contents in steel.

In the experiments, the mass percentage ratios of CaO/Al_2O_3 in slag varied in the range of 1.0–1.06. Relationship between T[O] content in steel and CaO/Al_2O_3 ratios in slag was shown in Fig. 4.13. As can be seen, with the increase of CaO/Al_2O_3 ratios in slag, T[O] in steel indicated an increasing tendency.

During the slag-steel reactions, slag played an important role in regulating the chemistry of strong deoxidizing elements in steel, including [Ca], [Mg] and [Al], which in turn importantly influenced chemical compositions of inclusions during the

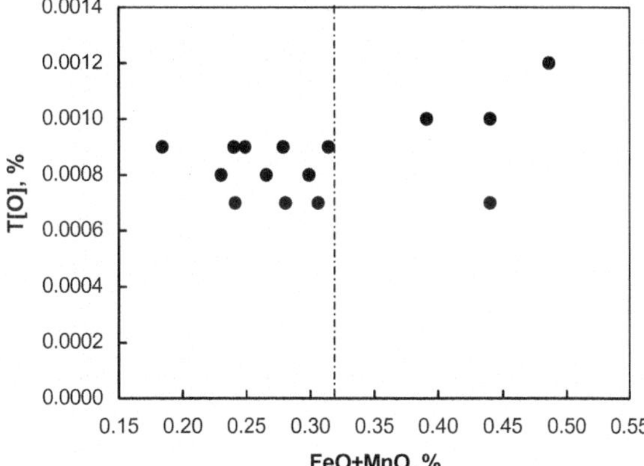

Fig. 4.12 Relation of (FeO + MnO) and T[O]

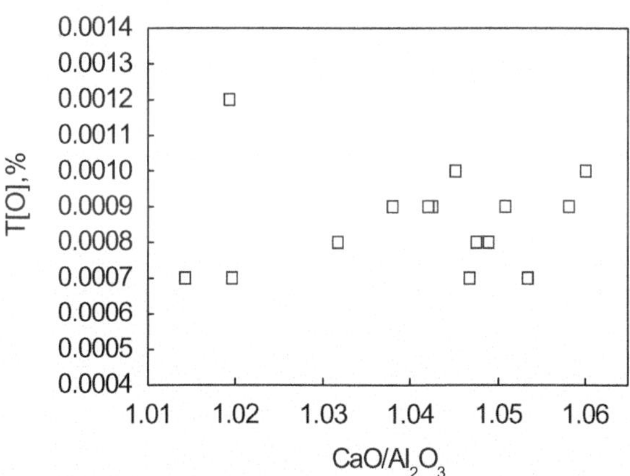

Fig. 4.13 Relationship between T[O] in steel and CaO/Al_2O_3 in slag

chemical reactions of steel-inclusion. As can be seen from Table 4.4, the contents of (CaO + MgO) in slag were about 51–54% after 90 min' reaction of slag-steel. It can be seen that the [Ca] contents in molten steel were in the range of 0.0005%–0.0012% while the [Mg] content were in the range of 0.0010%–0.0014%, as shown in Fig. 4.14.

Under strong Al deoxidation, [Al] in steel would react with refining slag. Considering that, relation between [Al] and [Mg] and between [Al] and [Ca] in steel were evaluated. It can be seen that with the increase of [Al] contents in steel, the contents of [Mg] in steel were in the scope of 0.0010%–0.0014% and indicated an increasing tendency. By comparison, contents of [Ca] were about 0.0005%–0.0009% while did not change so obviously, as shown in Fig. 4.15.

Relationship between CaO and MgO contents in slag and [Ca] and [Mg] contents in steel were also analyzed. It can be seen that [Ca] and [Mg] contents in steel slightly changed with the small fluctuations of CaO contents and MgO contents in slag, as shown in Figs. 4.16 and 4.17.

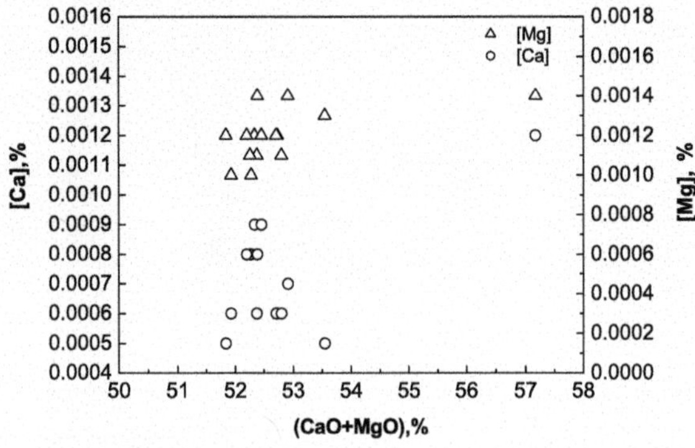

Fig. 4.14 Relationship between [Ca], [Mg] contents in molten steel and (CaO + MgO) contents in slag

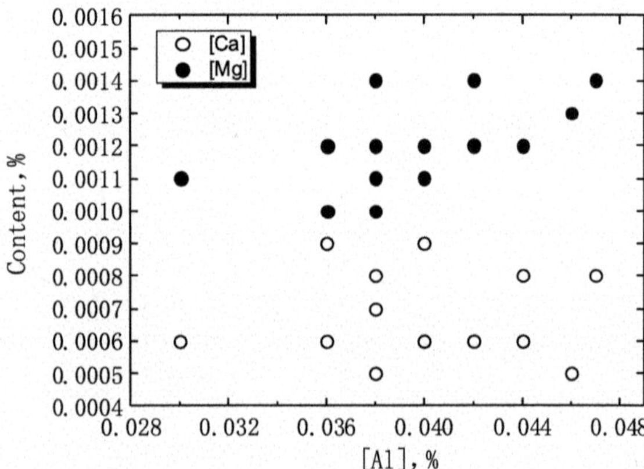

Fig. 4.15 Relationship between calcium, magnesium content and aluminum content in steel

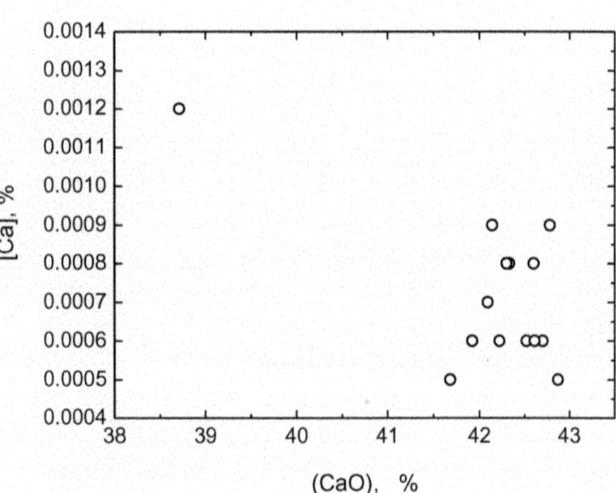

Fig. 4.16 Relation of CaO in slag on [Ca]

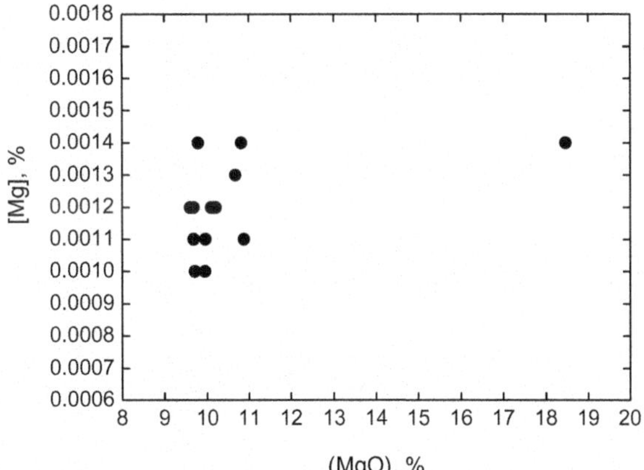

Fig. 4.17 Relation of MgO in slag on [Mg]

As it known, the [S] contents in steel were an important index to evaluate the cleanliness of steel. Higher basicity, low oxidizability of slag, low oxygen in molten steel were desirable for desulfurization. During the experiments, contents of [S] in steel varied from 0.0005% to 0.0007%, as shown in Figs. 4.18 and 4.19.

Because the [S] in steel was initially as low as 0.0047%, distribution ratios of sulfur ((%S)/[%S]) between slag and molten steel in experiments were mainly in the range of 26–30. As [Al] in steel and slag basicity in slag were in a relatively narrow range, influences of [Al] and slag basicity did not indicate obvious influence on (%S)/[%S], as shown in Figs. 4.20 and 4.21. As can be seen in Figs. 4.22 and 4.23, [S] contents in steel increased with the rise of (FeO + MnO) in slag while the ratio of (%S)/[%S] decreased with the rise of (FeO + MnO) contents in slag.

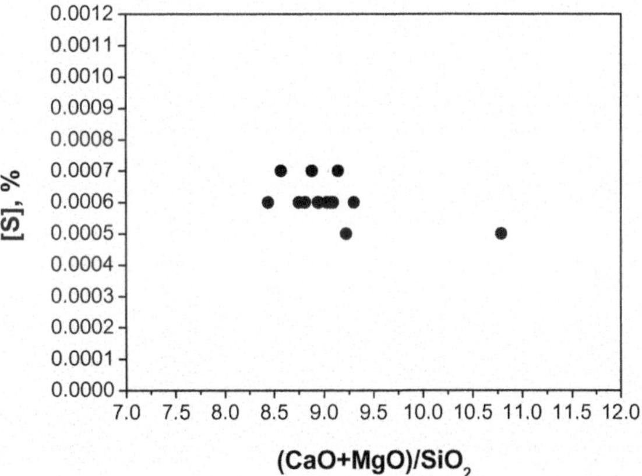

Fig. 4.18 Relation of slag basicity and [S] in steel

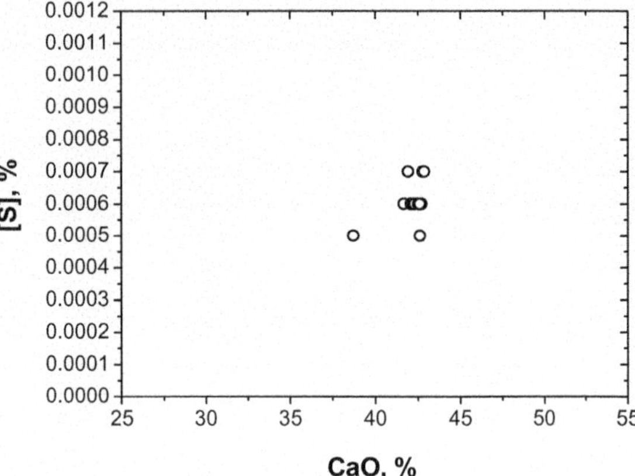

Fig. 4.19 Relation of CaO in slag and [S] in steel

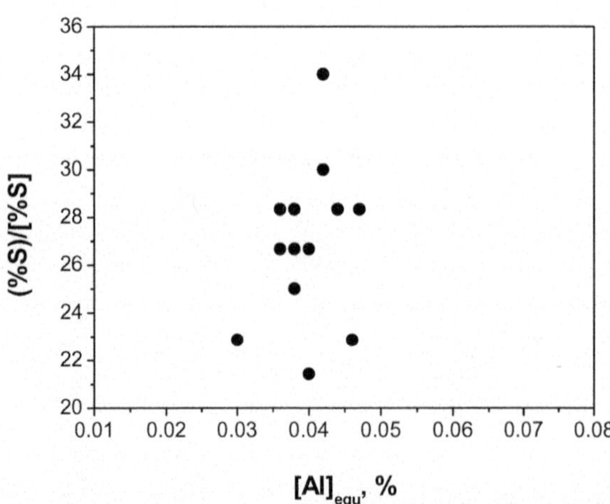

Fig. 4.20 Relation of (%S)/[%S] and [Al] in steel

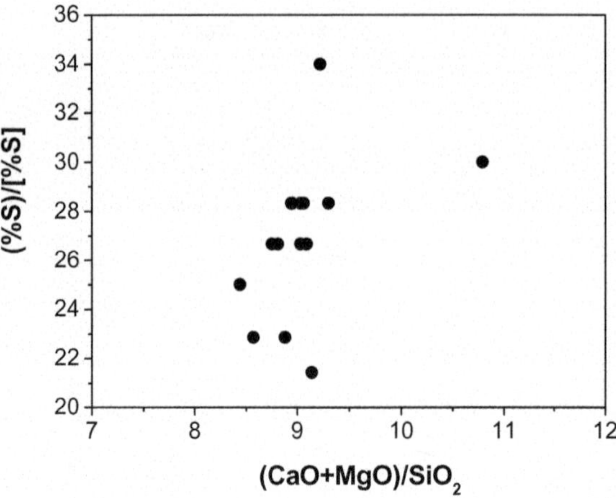

Fig. 4.21 Relation of (%S)/[%S] and slag basicity

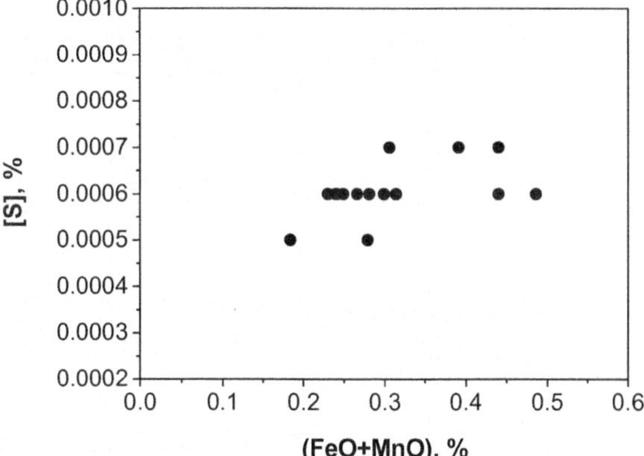

Fig. 4.22 Relation of [S] and (FeO + MnO)

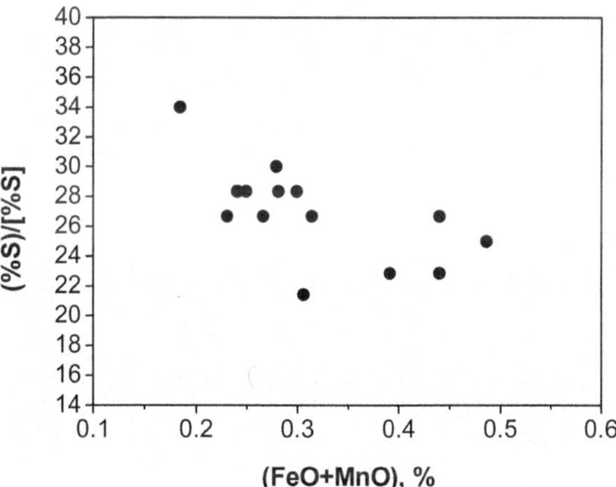

Fig. 4.23 Relation of (%S)/[%S] and (FeO + MnO)

4.3.2 *Non-metallic Inclusions in Steel*

The obtained steel samples were cut, ground, mirror-polished and afterwards observed under SEM to have information of inclusions, including type, size, morphology and chemical compositions etc. From SEM–EDS analysis results, it was found that inclusions in steel can be divided into two types according to their chemical compositions: (1) CaO–MgO–Al_2O_3–SiO_2 complex inclusions, some of which were with a few sulfur contents; (2) CaO–MgO–Al_2O_3–SiO_2 complex inclusions with high MgO contents. Moreover, contents of SiO_2 in all inclusions were very limited.

Figure 4.24 showed the morphology of typical inclusions observed during SEM–EDS detection. It can be seen that all the inclusions in steel were in regularly spherical shape and the sizes of inclusions were mostly less than 5 μm. The following Table 4.7 presented chemical compositions of inclusions. It can be seen that contents of

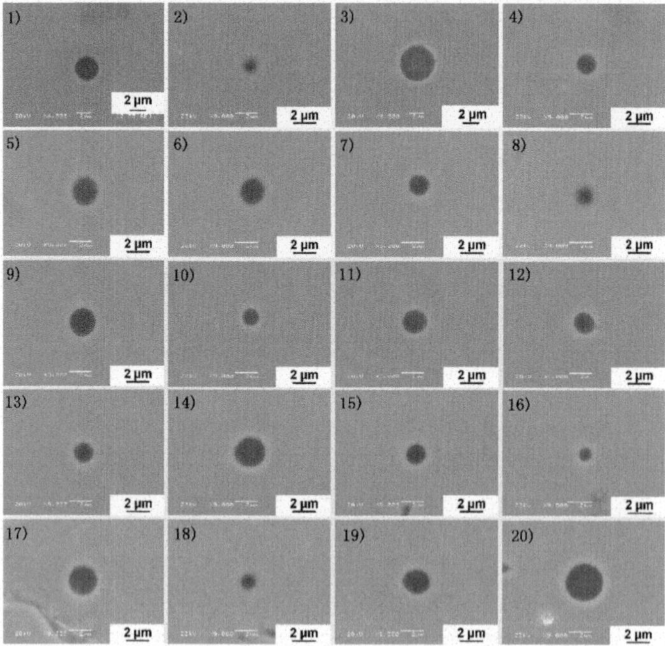

Fig. 4.24 Morphology of $CaO–SiO_2–Al_2O_3–MgO$ inclusions in steel

Table 4.7 Chemical composition of $CaO–SiO_2–Al_2O_3–MgO$ inclusions (mass%)

No.	Al_2O_3	SiO_2	CaO	MgO	Sum	No.	Al_2O_3	SiO_2	CaO	MgO	Sum
1	46.4	3.6	43.3	6.7	100	11	38.0	5.8	24.9	31.3	100
2	40.3	5.4	37.0	17.3	100	12	32.6	9.3	44.3	13.8	100
3	46.7	2.7	44.0	6.6	100	13	52.1	6.9	27.4	13.6	100
4	54.5	5.6	27.1	12.8	100	14	55.2	5.3	32.0	7.5	100
5	41.4	4.5	47.9	6.2	100	15	47.6	3.6	42.9	5.9	100
6	42.2	4.1	45.9	7.8	100	16	41.6	4.6	29.9	23.9	100
7	48.9	4.6	31.1	15.4	100	17	56.8	6.7	23.2	13.3	100
8	33.5	4.0	48.5	14.0	100	18	51.4	4.0	31.1	13.4	100
9	48.7	4.9	32.4	14.0	100	19	41.4	3.4	46.7	8.5	100
10	55.4	8.5	23.3	12.9	100	20	43.4	4.6	40.9	11.1	100

SiO_2 in inclusions were mainly below 5%. The contents of MgO in most inclusions were less than 20% while the contents of MgO in some inclusions exceeded 30%. Moreover, the contents of Al_2O_3 in inclusions were in the range of 40%–50%.

During the observation, there were about 500 inclusions were analyzed under SEM–EDS, of which 79.88% can be categorized as the first type while the other 20.12% were of the second type. The type, number fraction and size distributions

Table 4.8 Statistical results on the type and number of inclusions

Type		Ordinary $CaO–SiO_2–Al_2O_3–MgO$		High MgO–contained $CaO–SiO_2–Al_2O_3–MgO$
		Free of S	S-contained	
Index	Number	344	101	55
	Fraction, %	68.8	20.2	11.0

Table 4.9 Statistical analysis on the sizes of inclusion

Size	< 1μm	1–2 μm	2–3 μm	3–5 μm	5–7 μm
Number	28	271	133	60	6
Fraction, %	5.62	54.42	26.71	12.05	1.20

of the observed inclusions were shown in Tables 4.8 and 4.9, Figs. 4.25 and 4.26. It can be seen that inclusions smaller than 2 μm accounted for a number fraction of 60.04%, while the proportion of inclusions in the range of 2–3 μm was about 26.71%. Only 1.20% of the inclusions were with size greater than 5 μm.

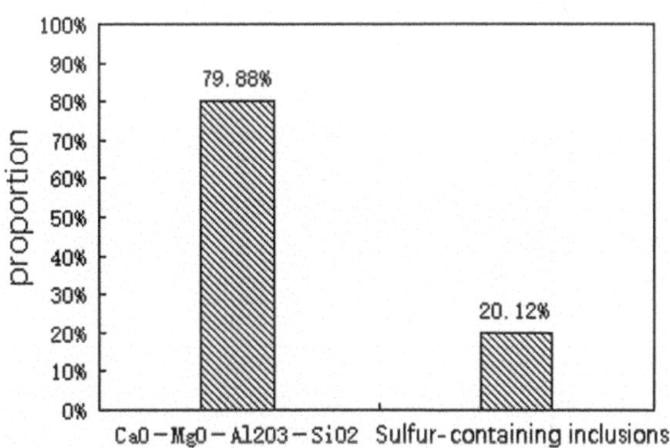

Fig. 4.25 Proportion of different types of inclusions

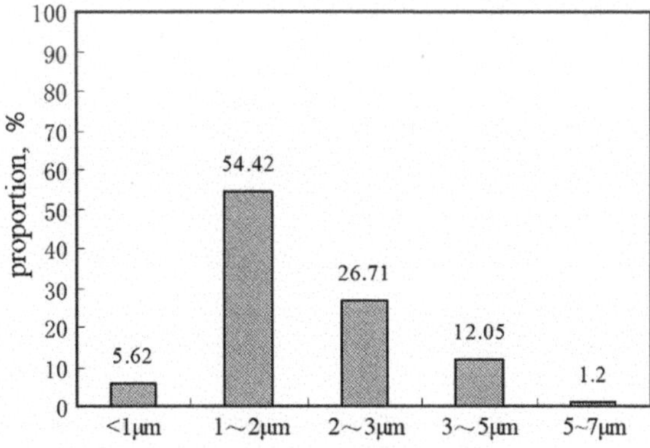

Fig. 4.26 Size districution of inclusions

According to the Stokes Law, floatation velocity of inclusions in liquid steel with diameter less than 150 μm can be expressed as following.

$$u_0 = \frac{(\rho_m - \rho_{in})g \cdot d_{in}^2}{18\eta_m} \tag{4.1}$$

In equation, u_0 was the floating velocity of inclusions (cm/s); ρ_m was the density of liquid steel, about 7.0 g/cm^3 at 1873 K; ρ_{in} was the density of inclusions, 3.5 g/cm^3 at 1873 K; d_{in} was the diameter of inclusions (cm); η_m was the viscosity of liquid steel, 0.03 g/(cm·s) at 1873 K; g was the acceleration of gravity 980 dyn/g. Therefore, according to equation, the time required for inclusions to float from the bottom of liquid steel to the interface between liquid steel and slag can be calculated as following.

$$t = \frac{h_m}{u_0} = 5 \times \frac{18\eta_m}{(\rho_m - \rho_{in})g d_{in}^2} \tag{4.2}$$

In the experiments, inner diameter of MgO crucibles was about 30 mm and the mass of molten steel was about 100 g. As the density of steel was about 7.0 g/cm^3 at 1873 K, depth of steel bulk was estimated about 20 mm. Hence, time required for inclusions of different sizes to float from the bottom of molten steel to the interface between molten steel and slag can be calculated. The relationship of inclusion size and floating time can be obtained and shown in Fig. 4.27. As can be seen, inclusions larger than 2 μm in diameter can float to the interface of steel-slag within 90 min.

Commonly, there were three kinds of behaviors for inclusions at the interface of steel-slag: (1) entering into the slag and be removed from steel; (2) staying at the interface of steel-slag; (3) returning to steel. Hence, although Stokes law calculation showed that inclusions about 2–10 μm have enough time to float up to the steel-slag interface within 90 min, inclusions with sizes in this range can still be observed in steel after 90 min of slag-steel reaction.

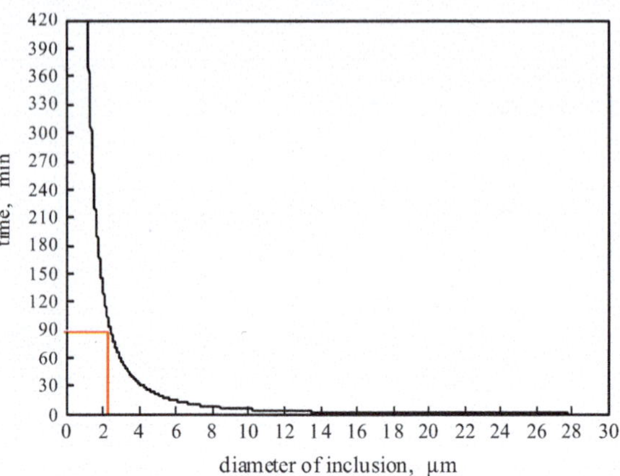

Fig. 4.27 Needed floatation time of inclusions in different sizes in the experiments

Figures 4.28, 4.29 and 4.30 showed the chemical composition distribution of inclusions in the (MgO + CaO)–Al_2O_3–SiO_2 pseudo-ternary system. As can be seen, when the equilibrium [Al] contents in liquid steel were in the range of 0.030%–0.038%, compositions of calcium magnesia aluminate inclusions in steel were well controlled inside or neighbor to the lower melting point region (1773 K) and contents of SiO_2 in inclusions were below 10%. When equilibrium [Al] contents were about 0.040%, compositions of inclusions contcentrated in the low melting point zone. With the rise of [Al] to about 0.040%–0.047%, compositions of inclusions in steel all located in the lower melting point region and its adjacent region.

As mentioned above, about 11% of the observed inclusions contained higher MgO contents. As SiO_2 contents in inclusions were less than 8%, in order to characterize

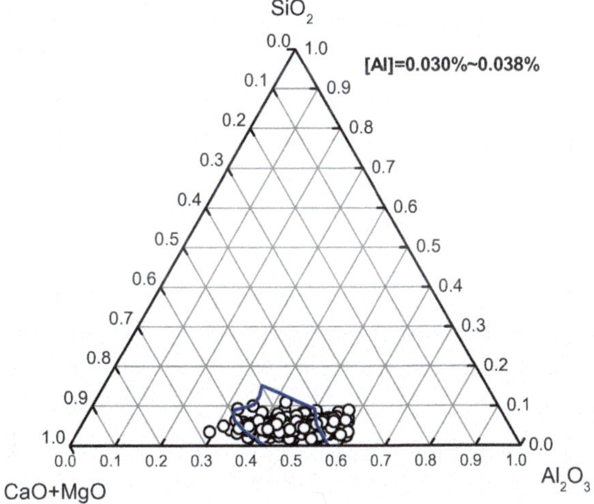

Fig. 4.28 Chemistry of inclusions with [Al] = 0.030–0.038%

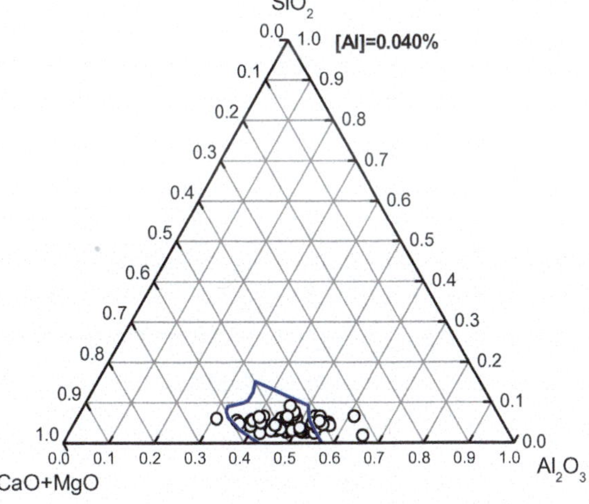

Fig. 4.29 Chemistry of inclusions: [Al] = 0.040%

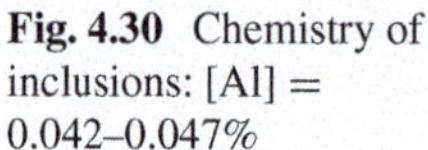

Fig. 4.30 Chemistry of inclusions: [Al] = 0.042–0.047%

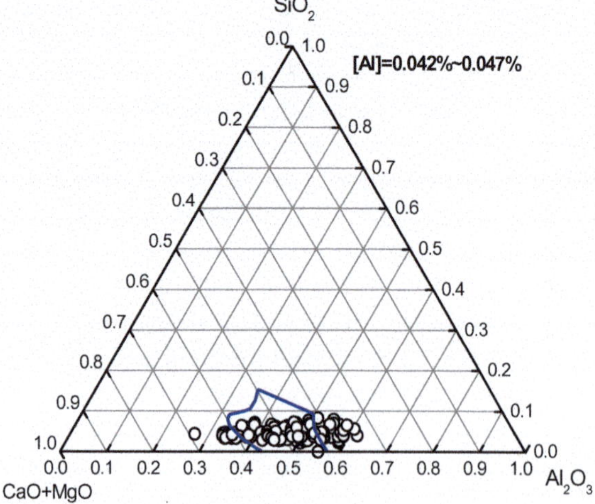

clearly the featues of MgO contents inclusions, low melting point region (<1773 K) in the phase diagram of CaO–MgO–Al_2O_3 pseudo-ternary system was calculated by the factsage software, and compositions of all the observed inclusions were projected into this pseudo-ternary system, as shown in Fig. 4.31. Figure 4.32. It can be seen that under the refining of slag with basicity about 8 and Al_2O_3 content about 45%, average compositions of inclusions in each heat were well controlled in the low melting point region or its adjacent region.

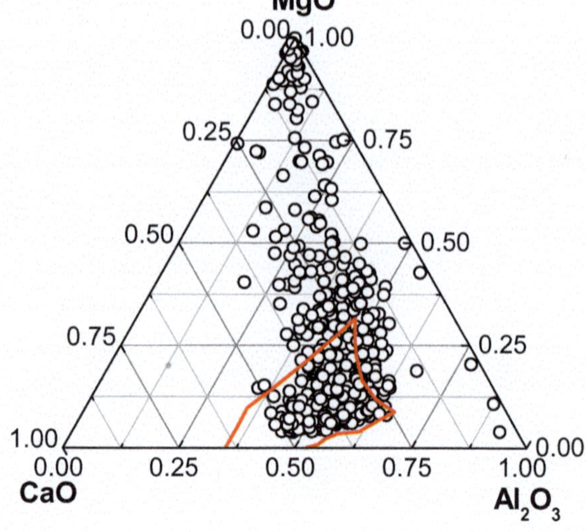

Fig. 4.31 Compositions of inclusion under slag A, reprinted from a previous paper of authors in Steel Research International, 2010, vol. 81, pp. 759–765, with permission from John Wiley and Sons

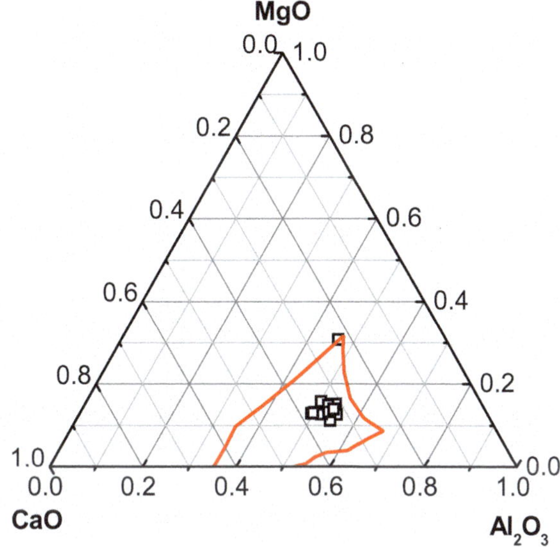

Fig. 4.32 Average composition of inclusion of each heat in the experiments

4.4 Control of Inclusions in Steel Refined by High Basicity High Alumina Slag B

4.4.1 Chemical Compositions of Steel Samples and Slag Samples

As stated in the Sect. 4.1, two types of high basicity high alumina slags were used in the study. The slag A was initially with about 45% Al_2O_3 while the slag B contained about 35%. The aim was to elucidat the influence of Al_2O_3 contents in slag on the control of inclusions in Al deoxidized steel. In the chemical reaction experiments between steel and slag B, chemical compositions of obtained steel samples and slag samples after 90 min' reaction were shown in Tables 4.10 and 4.11.

As can be seen, contents of [C], [Si], [Mn], [S] and [Cr] in steel were in the range of 0.40%–0.49%, 1.36%–1.73%, 0.81%–0.82%, 0.0003%–0.0008% and about 0.18%, respectively. While the contents of T[O] and [N] were in the range of 0.0005%–0.0010% and 0.007%–0.0021%, respectively. While contents of [Al], [Ca] and [Mg] in steel were in the scope of 0.032%–0.053%, 0.0003%–0.0012% and 0.0012%–0.0020%, respectively.

As shown in Table 4.11, CaO, Al_2O_3, SiO_2, MgO, MgO and FeO in slag were about 49.3%–50.78%, 30.30%–33.65%, 6.80%–7.49%, 9.004%–4.73%, 0.018%–0.038% and 0.09%–0.36%, respectively.

After 90 min' slag-steel reaction, mass percentage ratios of (CaO + MgO)/SiO_2 in slag were in the range of 7.90–9.14, and the contents of Al_2O_3 were in the range of

Table 4.10 Chemical compositions of steel samples (mass%)

Heat	C	Si	Mn	Cr	S	Al	Ca	Mg	T[O]	N
1	0.47	1.36	0.82	0.18	0.0004	0.033	0.0004	0.0014	0.0005	0.0021
2	0.40	1.38	0.82	0.18	0.0004	0.034	0.0004	0.0014	0.0006	0.0010
3	0.47	1.45	0.82	0.18	0.0003	0.034	0.0003	0.0012	0.0007	0.0021
4	0.45	1.48	0.82	0.18	0.0004	0.036	0.0004	0.0014	0.0007	0.0016
5	0.48	1.64	0.82	0.18	0.0003	0.046	0.0005	0.0018	0.0007	0.0008
6	0.46	1.63	0.82	0.18	0.0004	0.053	0.0006	0.0017	0.0006	0.0006
7	0.47	1.66	0.82	0.18	0.0004	0.042	0.0010	0.0017	0.0006	0.0018
8	0.45	1.64	0.82	0.18	0.0005	0.040	0.0007	0.0018	0.0009	0.0016
9	0.49	1.78	0.82	0.18	0.0005	0.046	0.0006	0.0018	0.0009	0.0016
10	0.46	1.73	0.81	0.18	0.0006	0.036	0.0012	0.0018	0.0009	0.0016
11	0.47	1.23	0.80	0.17	0.0006	0.032	0.0006	0.0012	0.0010	0.0010
12	0.49	1.71	0.79	0.18	0.0008	0.039	0.0012	0.0020	0.0007	0.0007

Table 4.11 Chemical compositions of slag samples (mass%)

Heat	MgO	Al_2O_3	SiO_2	CaO	(%S)	MnO	FeO	MnO + FeO
1	9.78	31.26	7.41	50.68	0.014	0.022	0.09	0.112
2	9.22	31.53	7.44	50.78	0.016	0.023	0.09	0.113
3	9.82	31.94	7.35	49.91	0.015	0.021	0.09	0.111
4	9.29	31.89	7.49	50.63	0.016	0.025	0.07	0.095
5	9.5	33.65	6.84	49.3	0.014	0.02	0.08	0.100
6	9.24	31.83	6.97	49.91	0.015	0.021	0.09	0.111
7	9.73	31.9	7.16	50.38	0.013	0.018	0.12	0.138
8	14.73	30.3	6.8	47.45	0.016	0.024	0.09	0.114
9	9.22	31.32	7.43	50.9	0.014	0.024	0.36	0.384
10	9.43	32.33	7.08	49.35	0.015	0.038	0.09	0.128
11	9.23	32.92	7.52	50.21	0.016	0.04	0.16	0.200
12	9.00	33.04	7.46	50.52	0.012	0.02	0.16	0.180

30.30–33.65%. In the experiments, slag compositions were very stably controlled, as shown in Figs. 4.33 and 4.34.

Due to the erosion of MgO crucibles by slag, MgO in crucibles dissovled into slag and there were certain amounts of MgO in the slag of every heat. After 90 min of steel-slag reaction, MgO contents in slag were in the ranges of 9%–10%, as shown in Fig. 4.35. As it known, the contents of (FeO + MnO) in slag indicated the oxidizability of slag to liquid steel. During the experiments, contents of (FeO + MnO) and ratios of (CaO + MgO)/SiO_2 in slag B were shown in Fig. 4.36. As can be seen, contents of (FeO + MnO) in slag B were moslty below 0.15%.

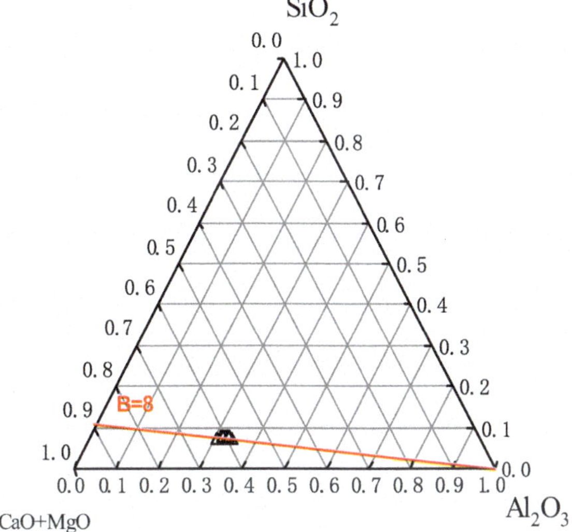

Fig. 4.33 Compositions of slag samples after the experiments (with the use of slag B)

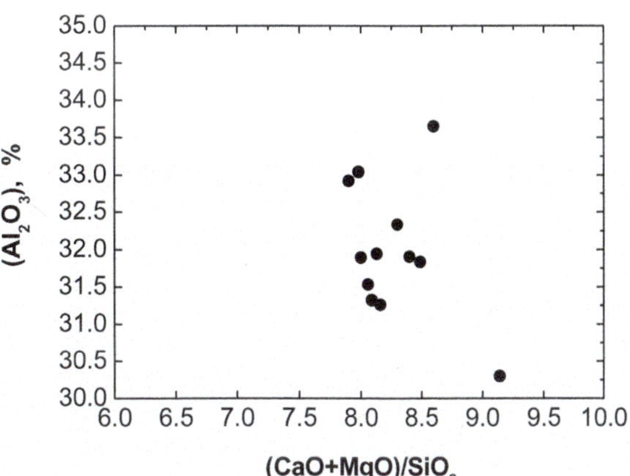

Fig. 4.34 (CaO + MgO)/SiO_2 and Al_2O_3 contents in slag B

During the experiments, initial [Al] contents in steel widely varied from 0.0050% to 0.0750%. Despite of great changes in initial [Al] contents, after 90 min' reaction between molten steel and slag, [Al] contents tended to approach a stable value and were in the range of 0.04%–0.055%, as shown in Fig. 4.37. This phenomenon in turn indicated that chemical reaction equilibrium between slag and steel was approximately established.

As it known, the contents of (FeO + MnO) in slags have important influences on the content of T[O] in steel. During the experiment, contents of (FeO + MnO) in slag of most heats were very low, ranging from 0.10% to 0.15% and only a few heats with (FeO + MnO) content of 0.40%. Moreover, with the increase of (FeO +

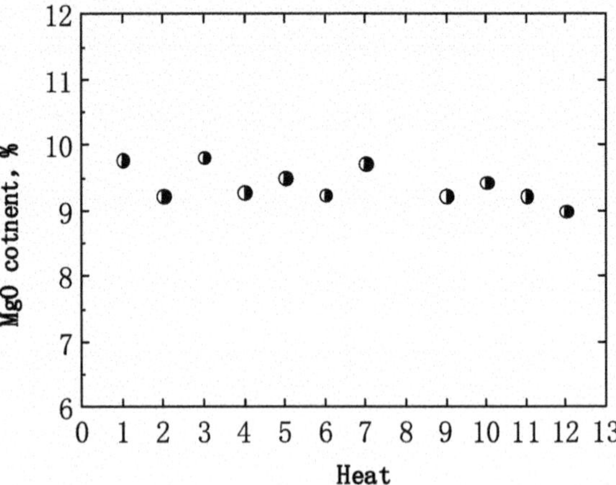

Fig. 4.35 Contents of MgO in slag B

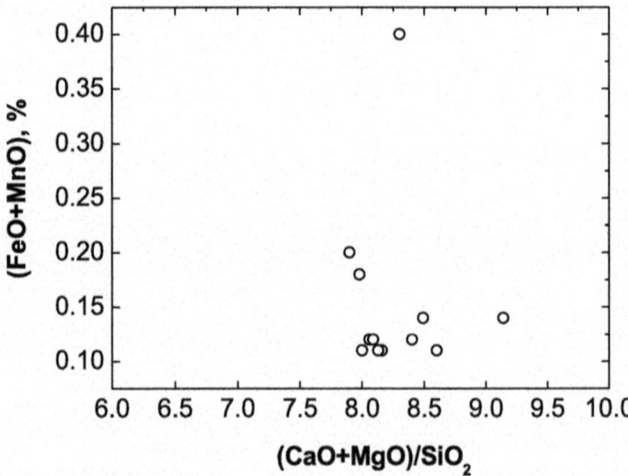

Fig. 4.36 (CaO + MgO)/SiO_2 ratios and (FeO + MnO) in slag B

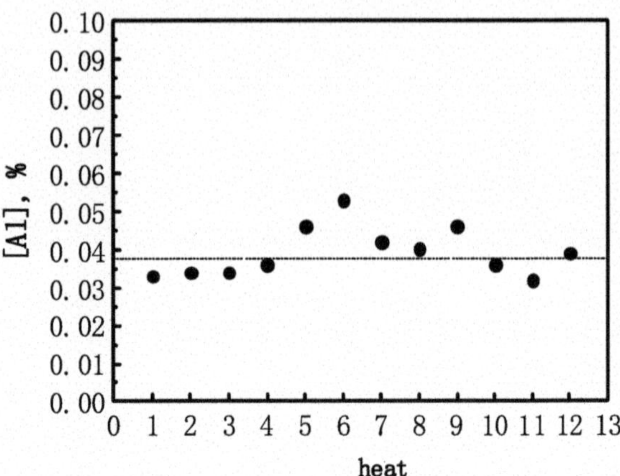

Fig. 4.37 Contents of [Al] in steel

MnO) contents in slag, T[O] contents in steel declined from 0.0007% to 0.0005%, as shown in Fig. 4.38.

Basicity of slag also greatly influenced T[O] contents in molten steel. Generally, the higher basicity of slag was, the lower T[O] contents would be contained in steel. During the experiments, the ratios of (CaO + MgO)/SiO_2 were in the scope of 7.90–9.14 while T[O] contents in steel were in the range of 0.0005–0.0009%. Moreover, with the increase of (CaO + MgO)/SiO_2, T[O] contents in steel decreased, as shown in Fig. 4.39.

Influences of CaO/Al_2O_3 ratios in slag on T[O] contents in steel were shown in Fig. 4.40. As can be seen, with the increase of CaO/Al_2O_3 from 1.4 to 1.7, T[O] contents in steel decreased.

As can be seen, with the refining of slag with basicity about 8 while the content of Al_2O_3 about 35%, the contents of [S] in steel were of very low level and varied in the scope of 0.0003%–0.0006%, as shown in Figs. 4.41 and 4.42.

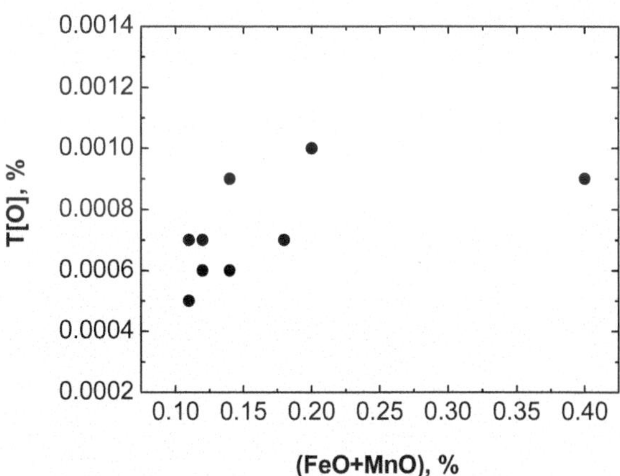

Fig. 4.38 Effect of (FeO + MnO) content on T[O]

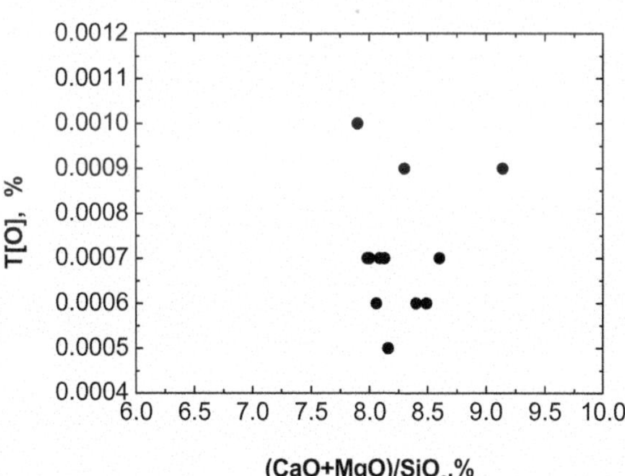

Fig. 4.39 Effect of (CaO + MgO)/SiO_2 on T[O]

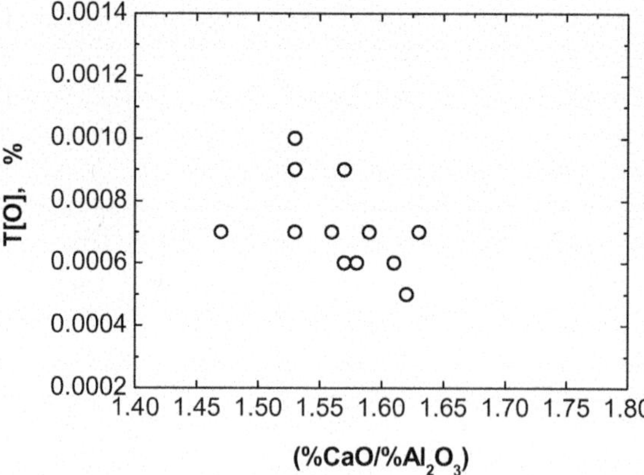

Fig. 4.40 Effect of CaO/Al_2O_3 on T[O] contents in steel

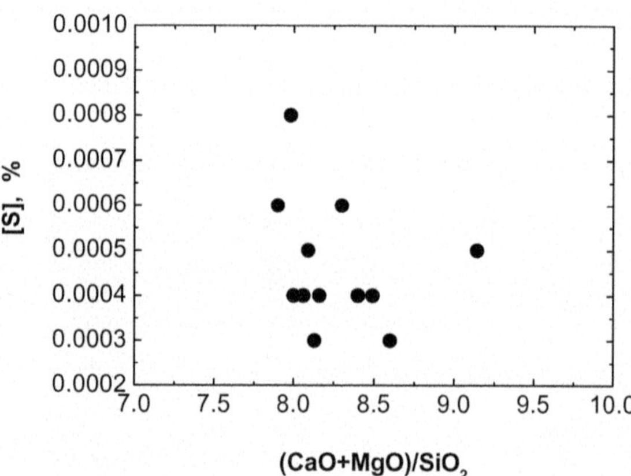

Fig. 4.41 Effect of $(CaO + MgO)/SiO_2$ on [S]

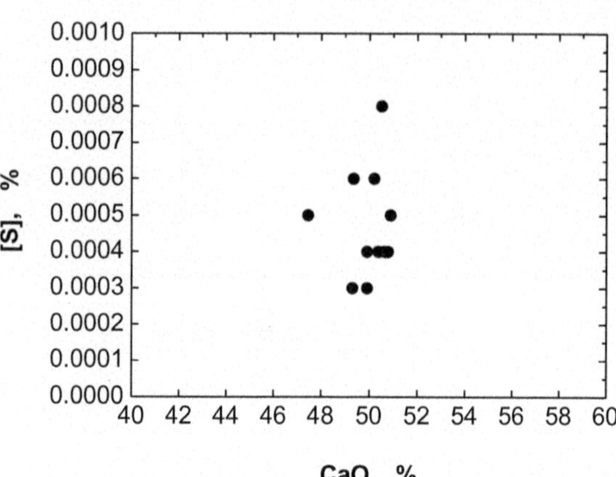

Fig. 4.42 Effect of CaO in slag on [S]

With the rise of slag (FeO + MnO) contents in the range of 0.1–0.5, contents of [S] in steel increased. It meant that slight fluctuations of (FeO + MnO) contents in slag in the range of 0.1%–0.5% can still influence the contents of [S] in steel, as shown in Fig. 4.43. As [Al] contents in molten steel were about 0.04%, [S] contents in steel did not indicate obvious relation to [Al], as shown in Fig. 4.44.

As strong deoxidizing elements including [Al], [Ca] and [Mg] etc. in steel greatly affect the control of inclusions in steel, contents of [Al], [Ca], [Mg] in molten steel after 90 min of reaction were analyzed and evaluated in details for the relationship between CaO–[Ca], MgO–[Mg], [Al]–[Ca] and [Al]–[Mg]. It can be seen in Fig. 4.45 that the [Ca] and [Mg] contents in steel varied from 0.0003% to 0.0012% and from 0.0012% to 0.0020%, respectively. With the increase of (CaO + MgO)/SiO_2 in slag, contents of [Ca] and [Mg] in molten steel also increased.

By comparison, the changes of [Ca] and [Mg] contents with the increase of CaO and MgO contents in slag were not so obvious, as shown in Figs. 4.46 and 4.47, respectively. In the experiments, effects of [Al] content on [Ca] and [Mg] content

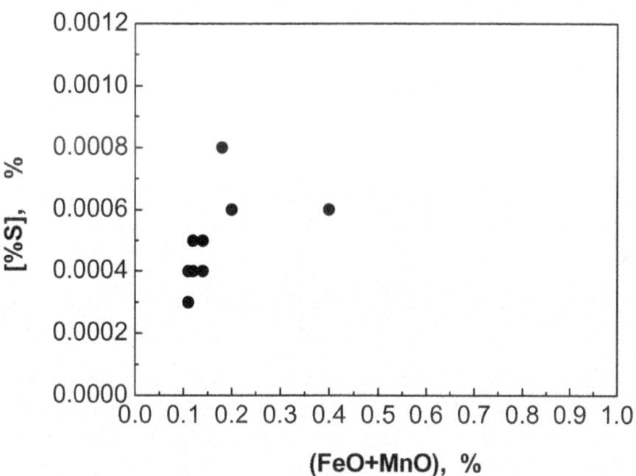

Fig. 4.43 Effect of (FeO + MnO) contents on [S]

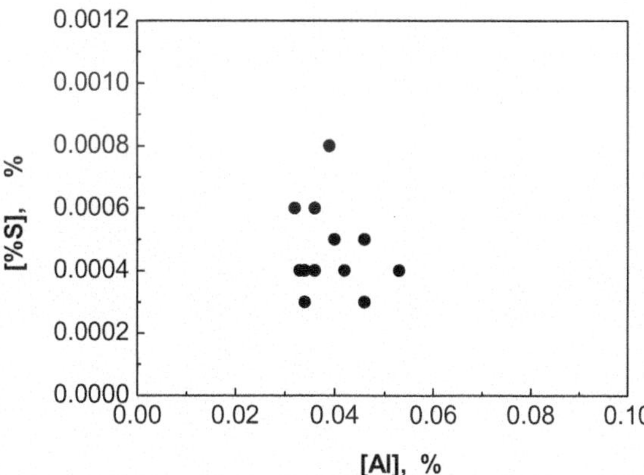

Fig. 4.44 Relation of [Al] and [S] in steel

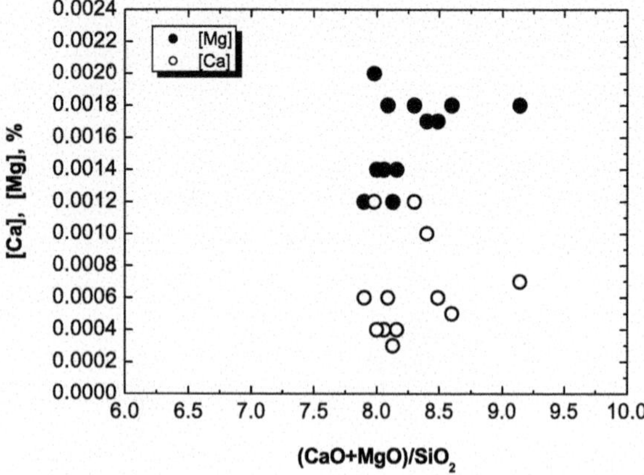

Fig. 4.45 Effect of $(CaO + MgO)/SiO_2$ in slag on [Ca] and [Mg]

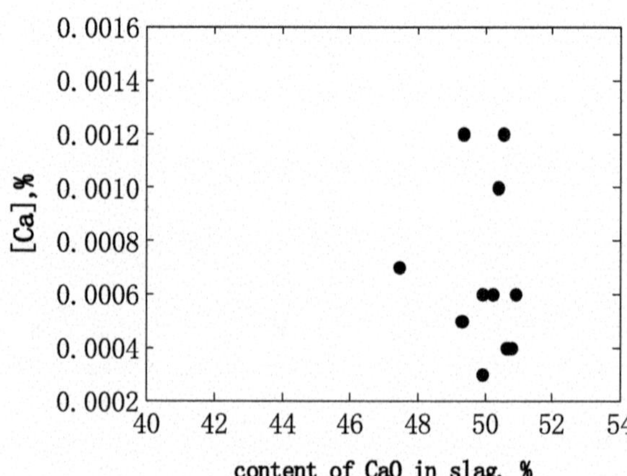

Fig. 4.46 Effect of CaO in slag on [Ca]

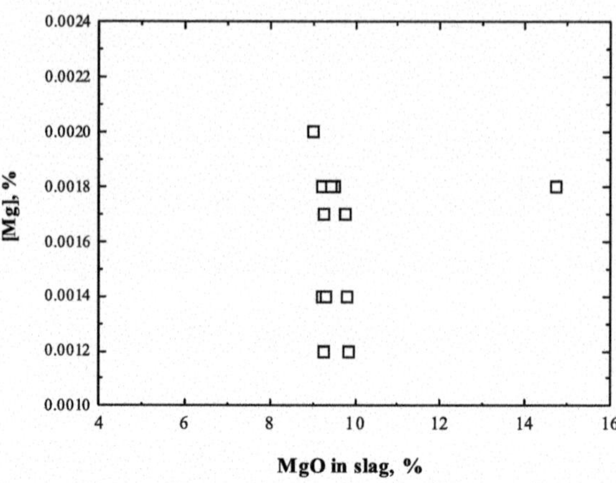

Fig. 4.47 Effect of MgO in slag on [Mg] in steel

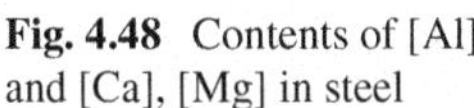

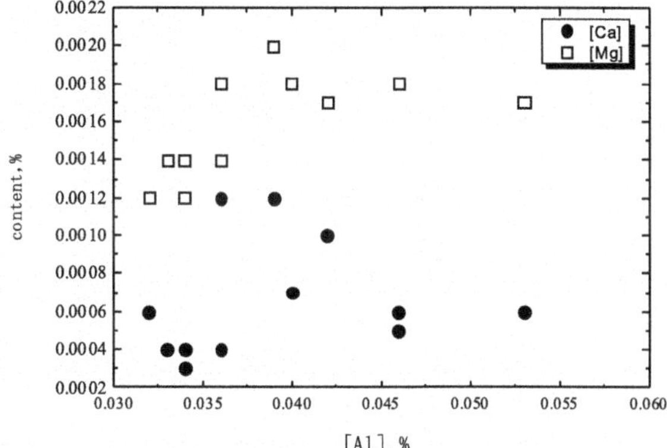

Fig. 4.48 Contents of [Al] and [Ca], [Mg] in steel

were more obvious: when [Al] content in steel increased from 0.033 to 0.053%, both the contents of [Mg] and [Ca] were increased from 0.0012% to 0.0018% and from 0.0003% to 0.0012% respectively, as shown in Fig. 4.48.

4.4.2 Non-metallic Inclusions in Steel

After cutting, grounding, mirror polishing, the steel samples were observed under SEM–EDS to obtain detailed information on the natures of inclusions, including their types, morphologies, sizes and chemical compositions etc.

It was found that the non-metallic inclusions in steel were mainly composed of CaO–MgO–Al_2O_3–SiO_2 system, with SiO_2 below 8%. According to the chemical composition, inclusions can be divided into the following two categories: (1) low melting point CaO–MgO–Al_2O_3–SiO_2 inclusions, which can be further classified into two kinds, viz. free of sulfur or with a few sulfur, and the latter were with very limited sulfides. (2) CaO–MgO–Al_2O_3–SiO_2 inclusions with much higher MgO content. In Fig. 4.49, typical morphology of CaO–MgO–Al_2O_3–SiO_2 inclusions in steel were indicated. It can be seen that the inclusions were in good spherical shape, with size less than 5 μm. Chemical compositions of these inclusions were indicated in Table 4.12, in which the contents of each component were provided.

Totally, 337 inclusions were observed in all the steel sample. It was found among these inclusions, 64.4% of them belonged to the first type while 35.6% were of the second type. The number and size of different kinds of inclusions are shown in Table 4.13 and Fig. 4.50, respectively.

It can be seen from Fig. 4.50 that the inclusions in steel are mainly composed of CaO–MgO–Al_2O_3–SiO_2 system composite inclusions, with a proportion of 64.4%. And most inclusions were free of sulfur. In addition, high MgO content inclusions were largely formed in steel, with a proportion as high as 35.6%.

It can be seen from Table 4.14 and Fig. 4.51 that inclusions with size less than 2 μm accounted for 72.1% of the total number of inclusions, while 22.6% of them were

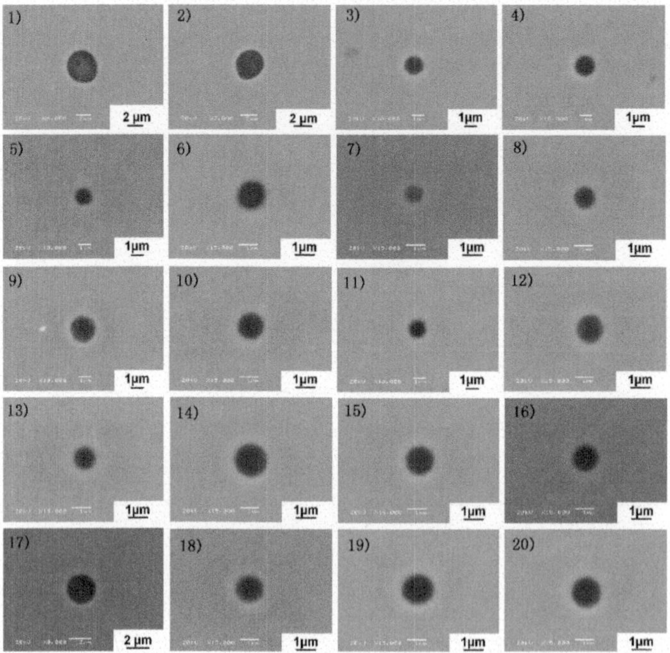

Fig. 4.49 Typical non-metallic inclusions in steel

Table 4.12 Chemical composition of CaO–MgO–Al_2O_3–SiO_2 inclusions (mass%)

No.	Al_2O_3	SiO_2	CaO	MgO	Sum	No.	Al_2O_3	SiO_2	CaO	MgO	Sum
1	43.6	7.1	30.2	19.1	100	11	45.3	8.1	36.7	9.9	100
2	44.6	6.5	43.3	5.6	100	12	49.0	6.8	36.0	8.2	100
3	43.1	5.4	44.0	7.5	100	13	50.2	7.3	33.5	9.0	100
4	10.4	3.6	46.5	39.5	100	14	47.4	7.3	37.3	8.0	100
5	31.9	11.2	17.1	39.8	100	15	35.1	7.3	40.3	17.3	100
6	32.1	7.2	23.5	37.2	100	16	47.0	7.5	31.6	13.9	100
7	44.1	8.4	32.6	14.9	100	17	31.0	5.0	28.4	35.6	100
8	31.9	6.0	29.5	32.6	100	18	49.3	6.6	36.0	8.1	100
9	40.5	6.4	49.4	3.7	100	19	33.9	5.7	56.3	4.1	100
10	47.4	9.8	30.9	11.9	100	20	28.8	5.5	42.7	23.0	100

Table 4.13 Statistical analysis of inclusions

Type	Ordinary calcium magnesia aluminate		High MgO content calcium magnesia aluminate
	Free of sulfur	With sulfur	
Number	197	20	120
Fraction, %	58.5	5.9	35.6

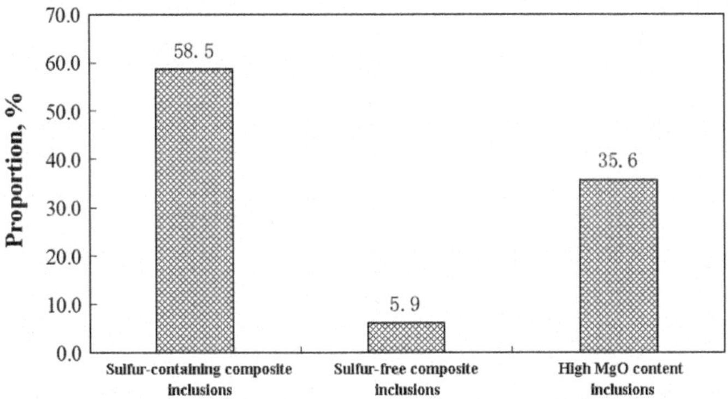

Fig. 4.50 Statistical analysis results of $CaO–MgO–Al_2O_3–SiO_2$ inclusions

Table 4.14 Size distributions of inclusions

Size	< 1 μm	1–2 μm	2–3 μm	3–5 μm	5–7 μm
Number	60	183	76	18	0
Fraction, %	17.8	54.3	22.6	5.3	0.0

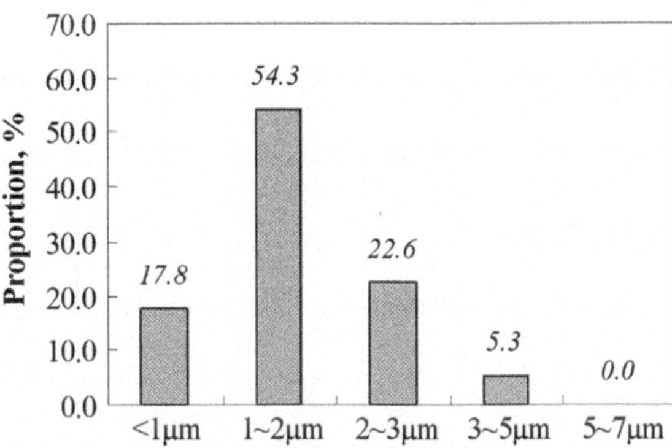

Fig. 4.51 Size distributions of inclusions

with sizes about 2–3 μm and only 5.3% of them were about 3–5 μm. Particularly, no inclusions with size greater than 5 μm are found.

Therefore, it can be seen that almost all inclusions in steel were less than 5 μm after 90 min' reaction. The reason for this was that the inclusion > 2 μm in liquid steel would be largely floated out, according to the prediction of Stokes formula, which has been discussion in previous section of this chapter.

Based on thermodynamic calculation in Chap. 3, it has been known that, the melting point of the inclusions would be very high when the MgO content is higher than 25%. Under this circumstance, the deformation ability of inclusions would be very poor, which would be very detrimental to the fatigue resistance property of

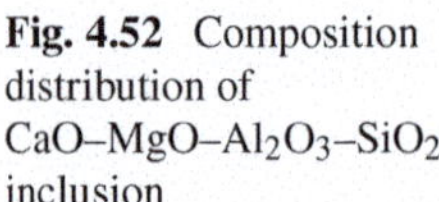

Fig. 4.52 Composition distribution of CaO–MgO–Al_2O_3–SiO_2 inclusion

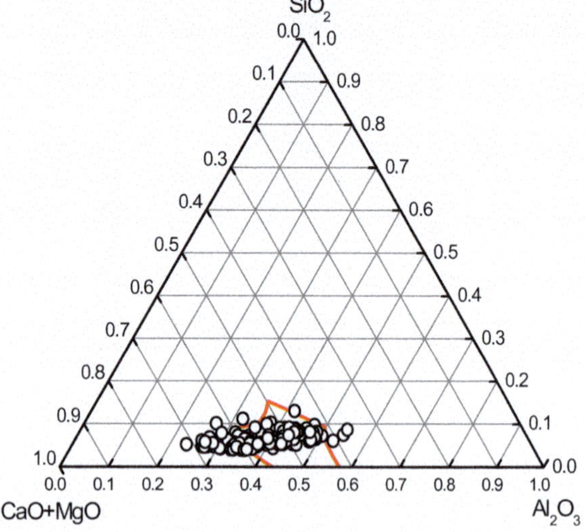

steel. As a result, in view of the inclusions with higher MgO contents in present study, 25% of MgO content was taken as a critical line to classify the inclusions as higher MgO content and lower MgO content. Excluding inclusions with high MgO content, inclusions with low MgO contents in the experiments were projected into the of (CaO + MgO)–Al_2O_3–SiO_2 pseudo-ternary, as shown in Fig. 4.52. The area surrounded by the blue solid line in the figure was the lower melting point area (with melting point below 1773 K). It can be seen that chemical compositions of inclusions were in the lower melting point region or the zone adjacent to the low melting point region.

For more reasonable and clearer characterization of the composition distributions of all the inclusions in steel, the average composition of CaO–MgO–Al_2O_3–SiO_2 complex inclusions (excluding inclusions with more than 40% MgO content) was projected into the pseudo ternary system of CaO–MgO–Al_2O_3 with 8% SiO_2 calculated by the FactSage software. The results showed that average composition of inclusions of most heats located in the lower melting point region, and located outside or at the boundary of the low melting point region for some heats, as shown in Fig. 4.53.

Average composition of the high MgO content inclusions were also analyzed and projected into the CaO–Al_2O_3–MgO ternary system with 8% SiO_2, as shown in Fig. 4.54. It can be seen that average composition of the High MgO contents CaO–MgO–Al_2O_3–SiO_2 inclusions deviated from the low melting point region and mainly located in the high melting point region.

Because the contents of SiO_2 in inclusions were less than 8% while the content of MgO in some inclusions were more than 40%, all the inclusions observed by SEM–EDS in the steel samples from each heat were projected into the CaO–MgO–Al_2O_3 pseudo-ternary system with 8% SiO_2. It was found that some of the inclusions

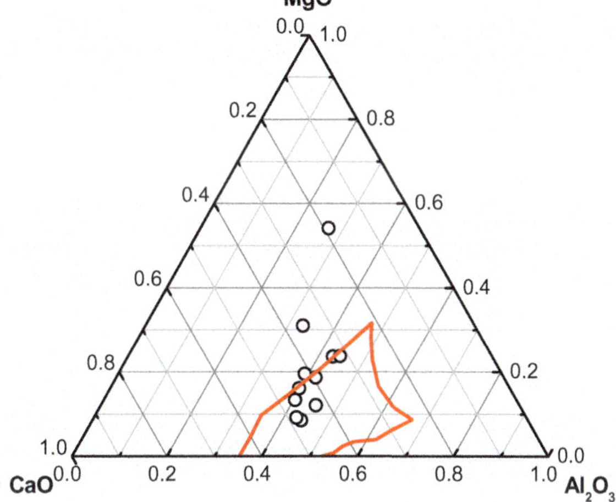

Fig. 4.53 Average composition of ordinary $CaO–MgO–Al_2O_3–SiO_2$ inclusions in all the heats

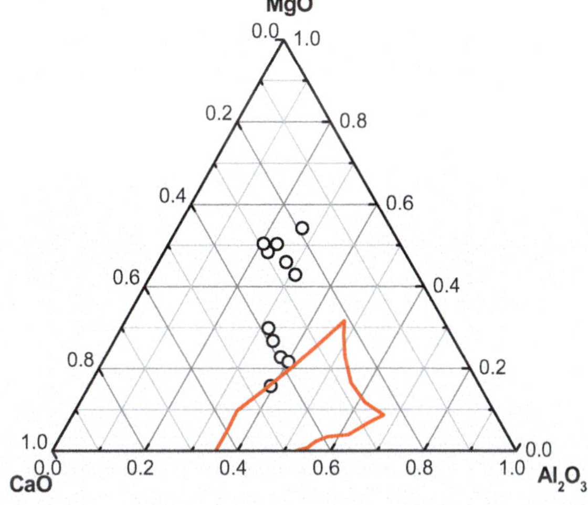

Fig. 4.54 Average composition of high MgO content inclusions in all the heats

located outside the lower melting point region while a most of them located in the high melting point region, as shown in Fig. 4.55.

As mentioned in the literature review in Chap. 1, Al, as the main deoxidizing element in molten steel, determines the chemical composition of deoxidizing products in steel. At the temperature of refining, [Al] in steel would react with slag to reduce CaO and MgO in slag, supplying [Ca] and [Mg] into molten steel. Subsequently, [Ca] and [Mg] would react with the existing deoxidized inclusions in the steel to promote the transformation in the chemical composition of inclusions.

Chemical composition of inclusions with [Al] contents about 0.032%–0.036%, 0.039%–0.042% and 0.046%–0.053% in the $CaO–MgO–Al_2O_3$ ternary system

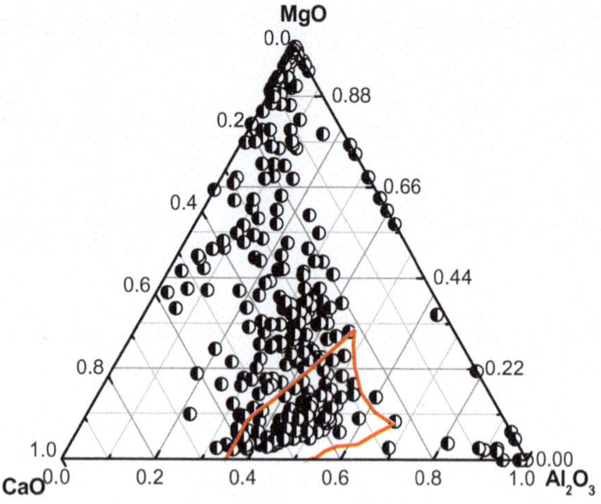

Fig. 4.55 Compositions of inclusion under slag B, reprinted from a previous paper of authors in Steel Research International, 2010, vol. 81, pp. 759–765, with permission from John Wiley and Sons

containing 8% SiO_2 was shown in Fig. 4.56a, b, c, respectively. It can be seen that with the increase of [Al] in steel, composition of inclusions slightly moved towards the corner of CaO. With the increase of [Al] content to 0.046%–0.053%, compositions of inclusions were relatively more concentrated in the lower melting point region. However, as can be seen in Fig. 4.56d, with the increase of [Al] contents, average composition of inclusions moved outside the low melting point region.

As it known, the melting point of inclusions mainly depends on their chemical compositions, which in turn would also affect the flotation behaviors of inclusions in steel. It has been proved that solid inclusions can be more difficult to wet in liquid steel thus can be easier removed while liquid or semi-liquid inclusions can be well wet by liquid steel thus would be more difficult to floated out. As [Al] showed influences on the compositions of inclusions, sizes of inclusions in steel melts with different level of [Al] contents were compared.

The results showed that at the end of slag-steel reaction, the proportion of inclusions below 1 μm decreased obviously with the increase of [Al] contents, while the proportion of inclusions within 1–3 μm increased with the rise of [Al] contents. This can be understood by the behaviors of solid inclusions in steel, which would easier to aggregate into large sizes. In the experiments, the proportion of inclusions in the range of 1–2 μm with different level of [Al] contents in steel were shown in Fig. 4.57.

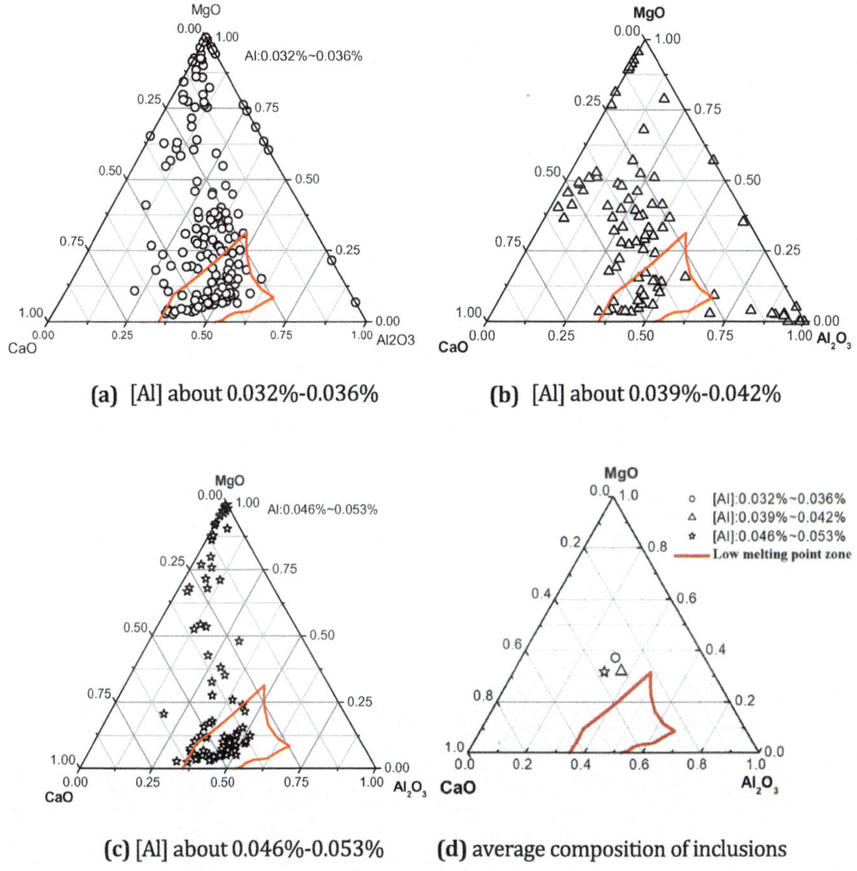

(a) [Al] about 0.032%-0.036%

(b) [Al] about 0.039%-0.042%

(c) [Al] about 0.046%-0.053%

(d) average composition of inclusions

Fig. 4.56 Effect of [Al] contents on the compositions of inclusions

4.5 Effects of Al_2O_3 Contents in High Basicity High Alumina Slag on the Control of Inclusions

As can be read in Sects. 4.3 and 4.4 in this chapter, it can be seen that the formed inclusions were similar but with different features. As a result, features of the inclusions were compared to analysis the influences of refining slags on the control of inclusions in Al deoxidized steel.

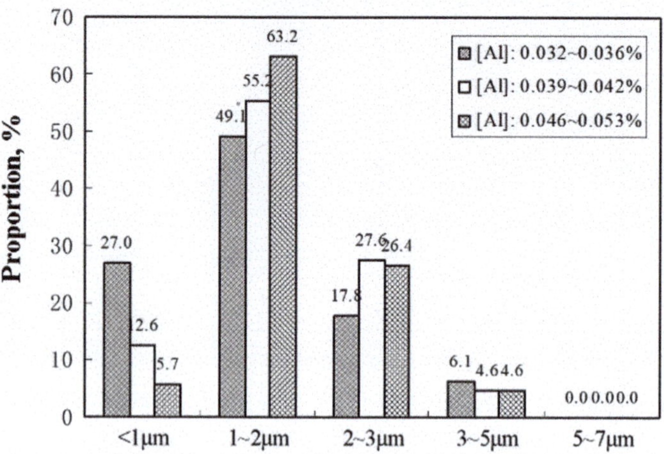

Fig. 4.57 Size distribution of inclusions in steel with different [Al] contents

4.5.1 Comparison on Chemical Compositions of Refining Slag

Chemical compositions of slag A and slag B of all the heats of experiments were projected to the (CaO + MgO)–Al_2O_3–SiO_2 pseudo-ternary phase diagram, as shown in Fig. 4.58. It can be seen that basicity of slag A and slag B was both about 8 at the end of reaction. In the process of slag-steel reaction, due to the increase of MgO, FeO, MnO and other components in slag, contents of Al_2O_3 in slag A was about 40%–42% while about 30%–33% in slag B after the reaction, with a difference is about 8%–10%.

It can be seen from Fig. 4.59 that the ratio of CaO/Al_2O_3 in slag A were about 1.0, while that in slag B were about 1.6. Whereas, the oxidizability of slag A was

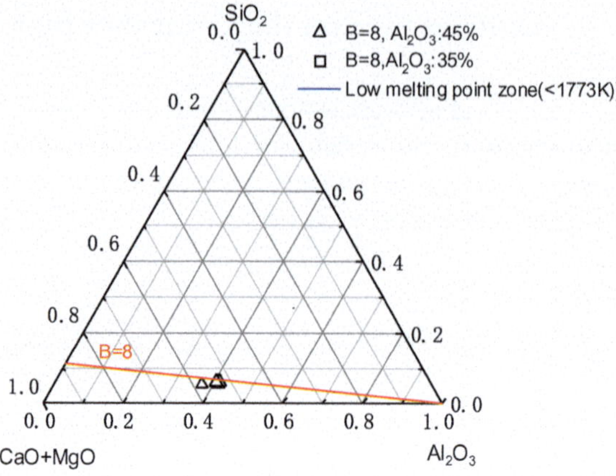

Fig. 4.58 Chemical compositions of slag A and slag B at the end of experiments

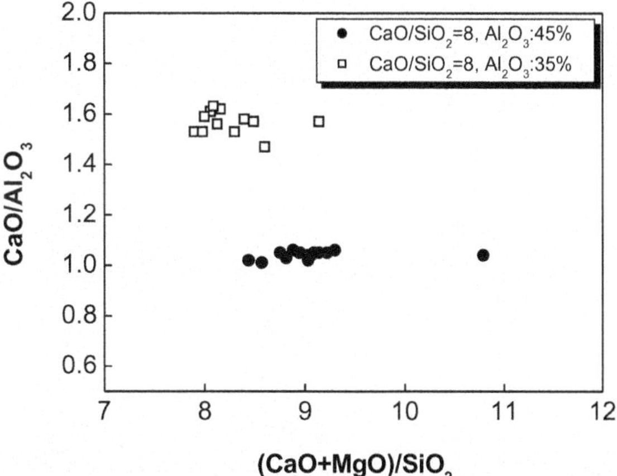

Fig. 4.59 CaO/Al_2O_3 ratios in slag A and slag B

higher than that of slag B, with the content of (FeO + MnO) about 0.2%–0.5% in the former while 0.1%–0.2% in the latter. Moreover, contents of (FeO + MnO) in slag A decreased obviously with the increase of $(CaO + MgO)/SiO_2$ while not obvious in slag B, as shown in Fig. 4.60.

Under the refining of slag A and slag B, the contents of MgO in slag were about 10% after 90 min of slag-steel reaction. With the use of slag A and slag B, the slag compositions at the end of the experiments were projected into the $CaO–Al_2O_3–SiO_2$ pseudo-ternary phase diagram with 10% MgO, as shown in Fig. 4.61. As can be seen, compositions of slag A located at the junction area of periclase and magnesia alumina spinel, with the melting point temperature lower than 1773 K. By comparison, chemical composition of slag B located in the zone with melting temperature higher than 1873 K.

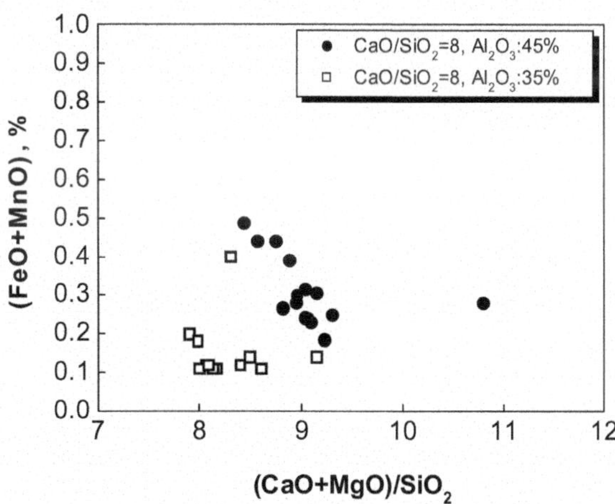

Fig. 4.60 Comparison of oxidizability of slag A and slag B after the experiments

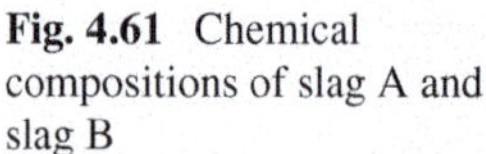

Fig. 4.61 Chemical compositions of slag A and slag B

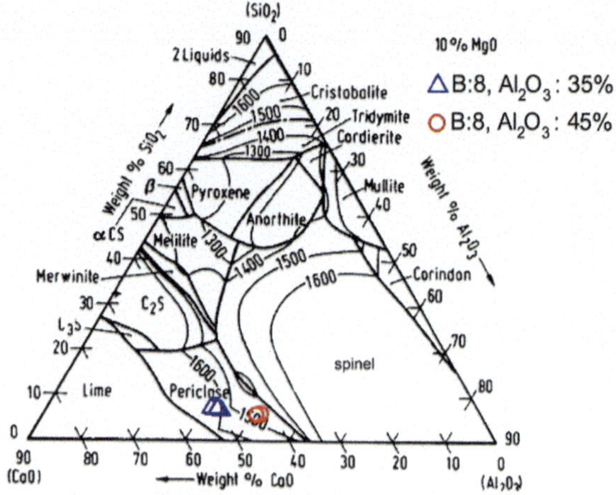

4.5.2 *Comparion on Steel Cleanness*

When Al strong deoxidization is used, the dissolved oxygen content in liquid steel would be reduced to very low content. As a result, T[O] in the steel can mainly represent the amounts of inclusions remained in steel. It meant that steel cleanness (content of oxygen and sulfur) would mainly depend on the ability of slag in absorbing inclusions and desulfurization. In present study, cleanness of steel refined by slag A and slag B was compared, as shown in Figs. 4.62 and 4.63. It can be seen that after 90 min' reaction, T[O] contents in steel refined by slag A and slag B were both concentrated in the range of 0.0007%–0.0010%. However, by comparison, T[O] contents in steel refined by the slag B was relatively lower. It can be seen from Fig. 4.63 that under the refining of slag A, the [S] contents in steel were between 0.0005% and 0.0007%. By comparison, under the refining of slag B, the [S] contents

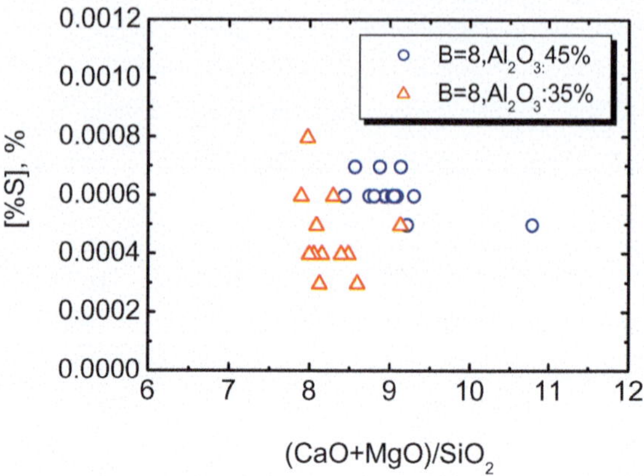

Fig. 4.62 T[O] in steel refined by slag A and B

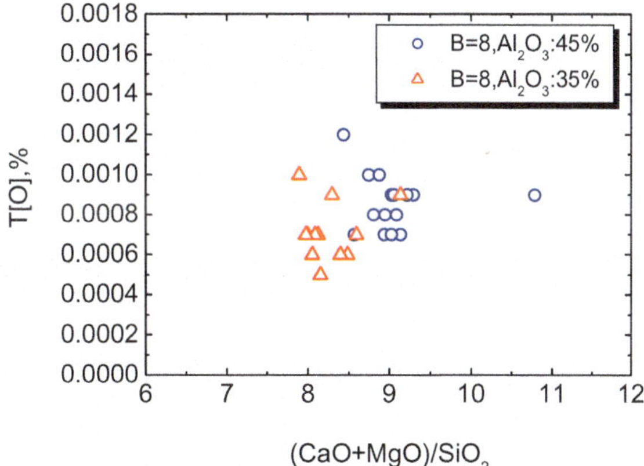

Fig. 4.63 [S] contents in steel refined by slag A and B

in steel were between 0.0003% and 0.0008%. It implied that higher CaO/Al_2O_3 ratios in the calcium aluminate slag would be favorable to reduce oxygen contents and sulfur contents in steel.

4.5.3 Comparison on Non-metallic Inclusions

According to the analysis of inclusion in Sects. 4.1 and 4.2, it can be known that the proportions of inclusions with high MgO contents in steel under the refining of slag A and slag B were obviously different. With the refining of the slag A, the number and proportion of inclusions with high MgO content were obviously lower than slag B. With the refining of steel under the slag B, proportion of inclusions with high MgO content in steel is 35.6%, while the ratio of high MgO content inclusions in steel were about 24% when the slag A was used. Differences in the chemical compositions of inclusions in steel melts refined by slag A and slag B can be read from Figs. 4.31 and 4.55.

As can be seen from Fig. 4.64, compared with slag B, the proportion of inclusions with sizes in the range of 3–5 μm was larger than slag A. While the proportion of inclusions about 1–2 μm was smaller and inclusions over 5 μm were found. By comparison, sizes of inclusions in steel were larger when slag A was used. The reason was that inclusions steel refined by slag A mainly located in the lower melting point region, existing as semi-liquid or liquid state in steel. Because of better wettability, they were harder to be removed. However, as stated above, more solid inclusions were produced in steel when slag B was used, which can be poorly wet by the liquid steel and thus can be easier to be floated out. Nevertheless, all the inclusions observed in the steel under the two slag conditions are spherical inclusions.

With the refining by slag A, the average composition of the inclusions in each heat was calculated. And the relationship between the average composition of inclusions and slag composition was analyzed. It was indicated that the ratio of (CaO +

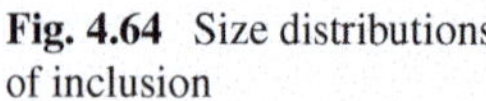
Fig. 4.64 Size distributions of inclusion

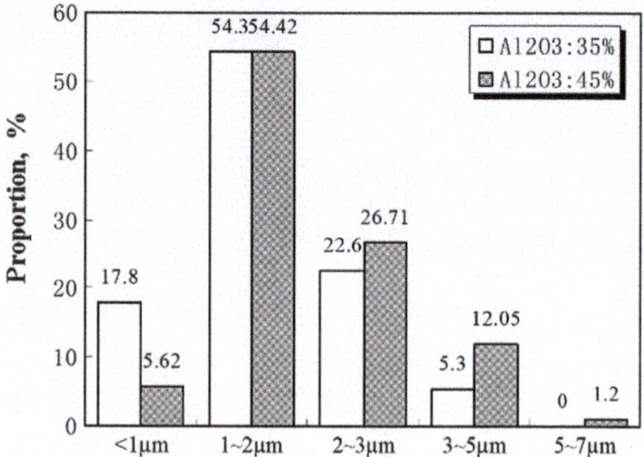

MgO)/SiO_2 in slag was very close to that in inclusions. However, the contents of Al_2O_3 in inclusions were higher than that in slag and the difference is between 5% and 10%, as shown in Figs. 4.65 and 4.66.

With the refining by slag B, average composition of CaO–MgO–Al_2O_3–SiO_2 system inclusions with MgO content below 40% in each heat were also projected into the pseudo-ternary (CaO + MgO)–Al_2O_3–SiO_2 system together with the slag composition, as given in Fig. 4.67. It can be seen that the ratios of (CaO + MgO)/SiO_2 in slags were equivalent to that of inclusions. While the contents of Al_2O_3 in inclusions were significantly higher than that in slag. The contents of Al_2O_3 in slag were about 30%–35% while that in inclusions were about 35%–40%. According to the

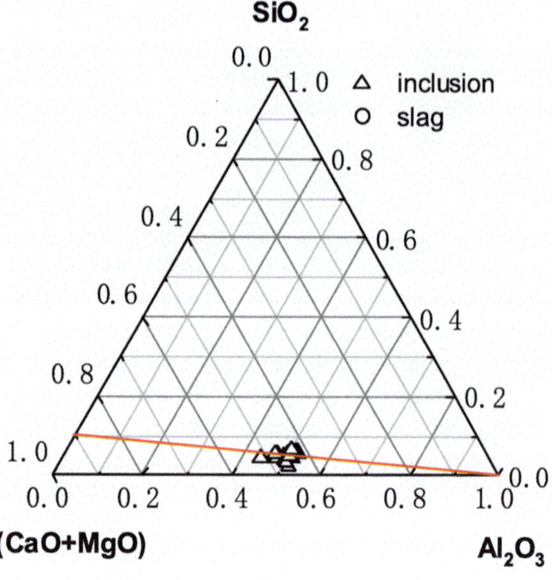

Fig. 4.65 (CaO + MgO)/SiO_2 in slag samples and inclusions in steel melts with the use of slag A

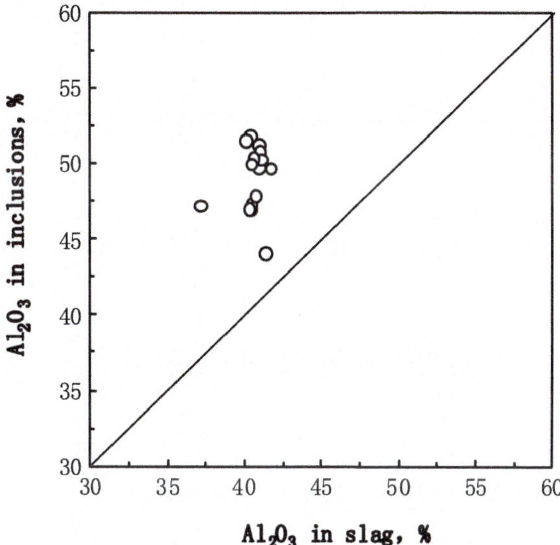

Fig. 4.66 Al_2O_3 contents in slag samples and inclusions in steel melts with the use of slag A

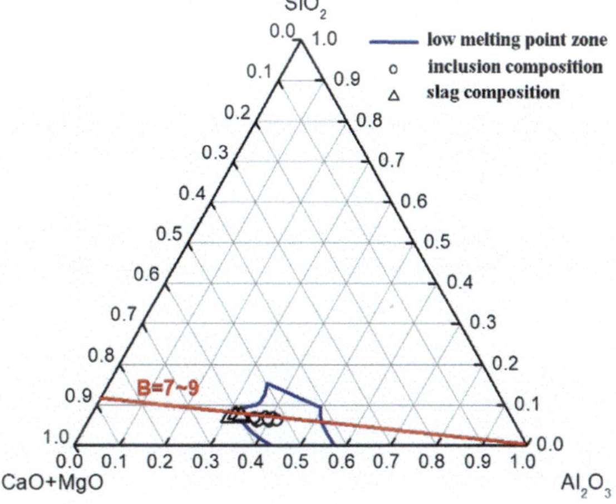

Fig. 4.67 (CaO + MgO)/SiO_2 in slag samples and inclusions in steel melts with the use of slag B

average composition of inclusions in each heat, relation between the content of Al_2O_3 in slag and the in inclusions were plotted at Fig. 4.68.

It can be seen that the ratios of (CaO + MgO)/SiO_2 in slag and inclusion were nearly equivalent, while the contents of Al_2O_3 in inclusion were always higher than that in slag. Theoretically, when complete chemical reaction equilibrium among slag, steel and inclusions were established, composition of inclusions should be identical to slag. However, the complete equilibrium was very difficult to approach although local equilibrium was possible.

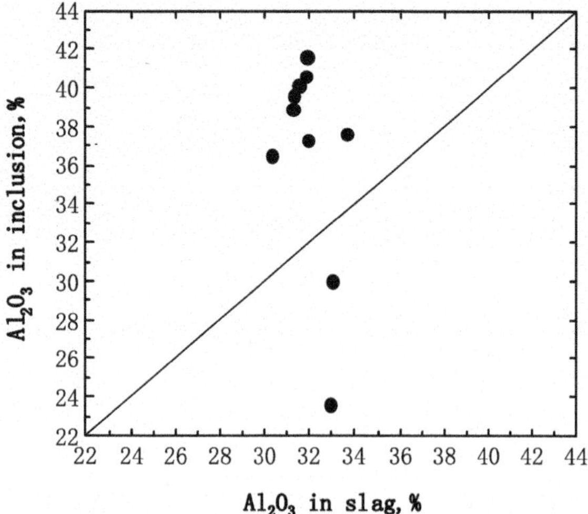

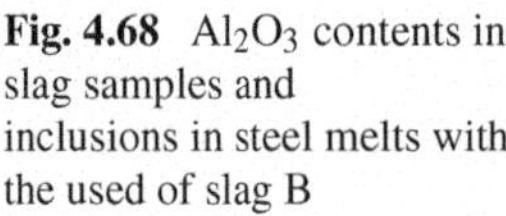
Fig. 4.68 Al_2O_3 contents in slag samples and inclusions in steel melts with the used of slag B

Different from slag, compositions of molten steel directly affected the chemistry of inclusions because of direct contact and chemical reaction between them. As mentioned in Chap. 1, as long as there were very small amounts of [Mg] and [Al] in the liquid steel, $MgO{\cdot}Al_2O_3$ inclusions can be easily formed. With the further increase of [Mg] contents in liquid steel, MgO inclusions would be produced. If there was a certain amount of [Ca] in the liquid steel, the liquid steel would react with $MgO{\cdot}Al_2O_3$ and MgO inclusions to transform them into calcium magnesium aluminate compound inclusions.

Therefore, to explain the formation of MgO-based calcium magnesium aluminates, the formation and transformation of $Al_2O_3/MgO{\cdot}Al_2O_3/MgO$ were discussed by thermodynamic calculations to obtain the stability phase diagram of them, as shown in Fig. 4.69. As can be seen, contents of [Mg] and [Al] in molten steel refined by slag B was mainly in the MgO formation region. While the contents of [Al] and [Mg] in molten steel mainly located in the zone of $MgO{\cdot}Al_2O_3$ or at the boundary of $MgO/MgO{\cdot}Al_2O_3$. The calculations revealed that inclusions in steel refined by slag B would be mainly MgO-based while the inclusions in steel refined by slag A would be mainly spinel together with some MgO-based ones. The calculations agreed well with experimental results.

Therefore, differences in the compositions of calcium magnesium aluminate inclusions in steel refined by slag A and slag B can be explained as follows:

(1) Under the refining by the slag A, contents of [Mg] and [Al] in steel were in the preferential formation region of $MgO{\cdot}Al_2O_3$ and MgO. The generated $MgO{\cdot}Al_2O_3$ and MgO would then continue to react with [Ca] in molten steel and were transferred into calcium magnesium aluminate inclusions with $MgO{\cdot}Al_2O_3$ and MgO as the core. Because some heats were with compositions locating at the boundary of $MgO{\cdot}Al_2O_3$ and MgO, therefore, some inclusions were with higher MgO content but were with lower proportion.

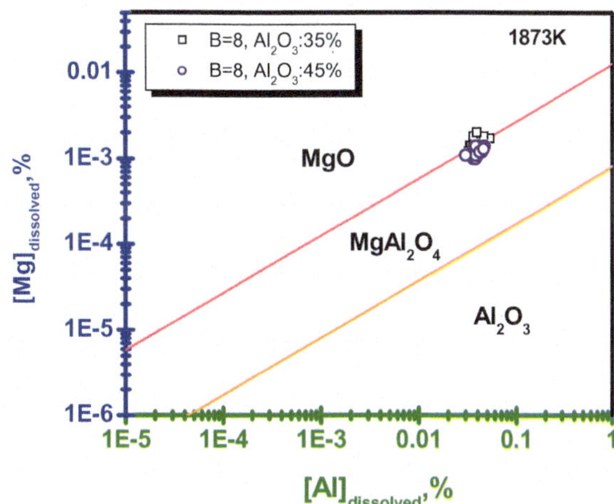

Fig. 4.69 Stability phase diagram of Al_2O_3/$MgO \cdot Al_2O_3$/MgO in steel

(2) Under the refining by the slag B, contents of [Mg] and [Al] in steel were in the preferential formation region of MgO. As a result, a large number of MgO inclusions in molten steel were formed. The generated MgO would then continue to react with [Al] and [Ca] in molten steel and were changed into calcium magnesium aluminate with higher MgO content.

4.6 Summary

In this chapter, chemical reaction between steel and high basicity high alumina slags were discussed in details, with emphasis on steel cleanness and control of non-metallic inclusions. Very important findings were obtained. Thermodynamic calculations in the previous Chap. 3 were verified that low melting point calcium magnesia aluminate inclusions can be targeted in steel with the used high basicity high alumina slag (with basicity about 8 while the alumina content about 35 and 45%). Moreover, the influences of alumina contents in the used refining slag system on steel cleanness and control of inclusions were also carefully compared. Based on the experimental results, conclusions can be drawn briefly as following.

(1) 90 min were reasonable for the slag-steel chemical reaction to approach local equilibrium. Despite of varied initial [Al] contents in steel, [Al] contents in steel melts changed towards the equilibrium content of [Al] predicted by thermodynamic calculation.

(2) Very high cleanness of steel can be obtained with the reining of the developed high basicity high alumina slag. T[O] contents in steel can be controlled within 0.0012% while the contents of [S] were limited in the range of 0.0003–0.0007%.

(3) Low melting point inclusions composed of calcium magnesia aluminate with SiO_2 below 8% were produced in the experiments. When the steel melts were

refined by slag A (with basicity about 8 while Al_2O_3 content about 45%), most of the inclusions concentrated in the low melting temperature zone (<1773 K). When the steel melts were refined by the slag B (with basicity about 8 while Al_2O_3 content about 35%), many inclusions located inside the low melting temperature zone (<1773 K) while there was also a large part of them located in the high MgO region in the pseudo-ternary system of CaO–MgO–Al_2O_3 with about 8%SiO_2.

(4) Compared to slag A, high MgO content inclusions can be much easier to produce in steel when slag B was used. Because chemical compositions of steel melts, viz. the contents of [Al] and [Mg], mainly located in the preferential formation zone of MgO. As a result, many MgO-based oxides would be formed besides spinel inclusions. Both of them would further react with liquid steel and be transferred into calcium magnesia aluminates in the end.

Chapter 5
Formation and Evolution Mechanisms of Inclusions in Steel

Abstract In steelmaking practice, time of refining is an issue closely related productivity and influence competitivity of steel company. During the study, reaction time is also found importantly affected the control of inclusions. During the reaction of steel and the used high basicity high alumina slags, as MgO-based refractory (MgO crucible) is used, inclusions experienced important evolutions from alumina to spinel and finally into low melting point $CaO–MgO–Al_2O_3$ with extended reaction time. In this chapter, the topic was discussed in details to elucidate the evolution mechanism of inclusions.

Based on the discussions in previous Chap. 4, it has been known that inclusions in the steel refined by slag A were mostly calcium magnesia aluminate with lower melting point. At the same time, a small amount of calcium magnesium aluminate inclusions with higher MgO content were also produced in steel.

As also mentioned in Chap. 1, even if the [Mg] and [Al] contents in molten steel were of very low level, Magnesia aluminate spinel inclusions can still be produced very easily. Spinel inclusions often have blocky shapes and with high melting points, which were very harmful to the fatigue resistance property of steel. When [Ca] entered into liquid steel, the spinel inclusions would be transformed into calcium magnesia aluminates. Hence, if the chemistry of steel can be controlled properly, low meting point inclusions with certain deformability in hot rolling can be well targeted with improved anti-fatigue performance of steel.

As previous studies on the formation and transformation of spinel to calcium magnesia aluminate inclusions mainly focused in the steelmaking field of stainless steel (such as SUS304 stainless steel), during which Si and Mn were used to deoxidize steel melt and the dissolved oxygen contents in molten steel would be higher. And the goals were usually to improve the surface quality of stainless steel while much less on fatigue property.

In order to produce spring steel with excellent anti-fatigue property, it is significant to study the formation and transformation of low melting point inclusions under Al

Part of the content in this chapter was reprinted from a previous publication of present authors in ISIJ international, vol. 50, pp. 95–104, copyright 2010, reprinted with the permission of ISIJ.

M. Jiang and X. Wang, *Slag-Steel Reaction and Control of Inclusions in Al Deoxidized Special Steel*, Engineering Materials, https://doi.org/10.1007/978-981-19-3463-6_5

deoxidation. However, this topic was seldom talked about. Therefore, this chapter attempted to explore the formation and transformation of inclusions in Al deoxidized spring steel under the refining of high basicity high alumina slag A in MgO crucible experiment by using laboratory equilibrium experiment. It should also note that high basicity high alumina slag was firstly proposed, developed and used by present authors for special steels.

5.1 Variations of Inclusions with Reaction Time

In the experiment, the initial composition of slag was featured by the binary basicity CaO/SiO_2 about 8, initial alumina content abpit 45%, and initial [al] content in steel about 0.0050%. In order to study the changes of inclusions with time at 1873 K, four heats of experiments were carried out, during which the slag-steel reaction time was 30 min, 60 min, 90 min and 180 min, respectively. At the end of each experiment, MgO crucible charged with liquid steel and slag was taken out for rapid water quenching. Then, the steel samples and slag samples were obtained and prepared for detailed chemical composition analysis and SEM–EDS inspection of inclusions. The compositions of molten steel and slag were shown in Tables 5.1 and 5.2 respectively.

5.1.1 Influence of Reaction Time on the Sizes of Inclusions

Non-metallic inclusions in steel were observed carefully under the SEM–EDS. Based on the result, statistical analysis on the size distributions of inclusions were obtained.

Table 5.1 Composition of steel samples at different reaction time (mass%)

Time (min)	C	Si	Mn	S	Cr	Al	Mg	Ca	T[O]
30	0.46	1.52	0.84	0.0005	0.17	0.038	0.0010	0.0003	0.0014
60	0.37	1.52	0.84	0.0006	0.17	0.041	0.0013	0.0002	0.0012
90	0.46	1.50	0.84	0.0008	0.17	0.042	0.0014	0.0005	0.0008
180	0.41	1.46	0.84	0.0005	0.17	0.036	0.0010	0.0005	0.0007

Table 5.2 Composition of slag samples at different reaction time (mass%)

Time (min)	CaO	Al_2O_3	SiO_2	MgO	MnO	FeO
30	44.90	40.50	6.90	6.56	0.045	0.72
60	44.30	40.13	6.26	8.61	0.038	0.26
90	42.68	39.37	6.95	9.67	0.042	0.24
180	42.43	39.47	6.57	10.90	0.048	0.48

The results showed that the proportion of large-scale inclusions in steel decreases obviously with the increase of reaction time. When the reaction time was 30 min, the proportion of inclusions in 3–5 μm was about 21.1%, while the fraction of inclusions in 5–7 μm was about 18.7%. With the increase of reaction time to 60 min, the proportion of inclusions about 3–5 μm in steel decreased to 16.7%, while the ratio of inclusions in 5–7 μm decreased. When the reaction time was increased to about 90 min, the proportion of inclusions with sizes about 3–5 μm decreased to 13.3%, and no inclusions were with sizes in the range of 5–7 μm. After 180 min' reaction, the ratio of inclusions with sizes below 1 μm was about 33.3% while the fraction of inclusions in the range of 1–2 μm was about 51.9%. Particularly, no inclusions were larger than 3 μm. As shown in Table 5.3 and Fig. 5.1, it can be seen that the size of inclusions in steel became ever smaller with the increase of reaction time.

In Figs. 5.2, 5.3, 5.4 and 5.5, morphology of typical inclusions in steel reacted with slag for 30 min, 60 min, 90 min and 180 min were indicated, respectively, chemical composition of the inclusions were shown in Tables 5.4, 5.5, 5.6 and 5.7, respectively.

Table 5.3 Proportion of inclusions with different sizes

Time (min)	< 1 μm (%)	1–2 μm (%)	2–3 μm (%)	3–5 μm	5–7 μm
30	7.9	31.6	21.1	21.1%	18.4%
60	8.3	25.0	36.1	16.7%	13.9%
90	3.3	36.7	46.7	13.3%	0
180	33.3	51.9	14.8	0	0

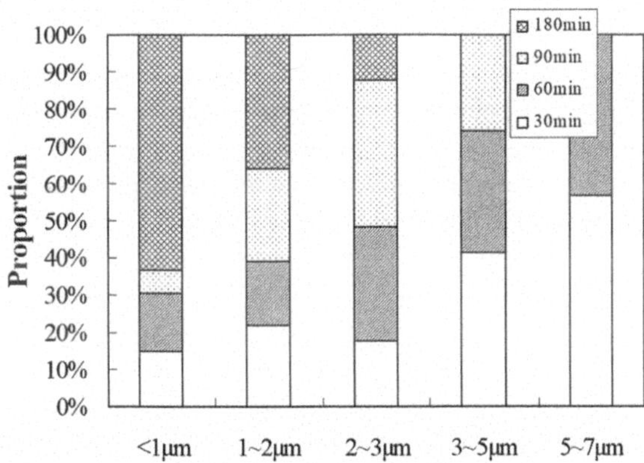

Fig. 5.1 Variations in the size of in steel with reaction time

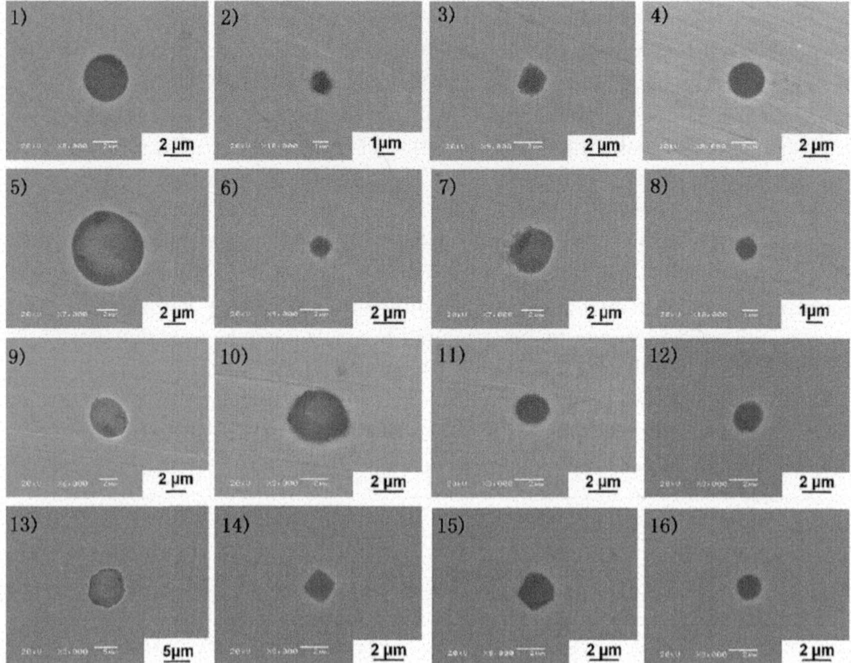

Fig. 5.2 Typical inclusion in molten steel reacted with slag for 30 min

5.1.2 Influences of Reaction Time on the Chemistry of Inclusions

Variations in the compositions of inclusion with reaction time can be clearer shown in Fig. 5.6. It can be seen that with the increased of reaction time from 30 to 60 min, the contents of Al_2O_3 in inclusions decreased from more than 50% to less than 40%. While the contents of MgO increased from about 30% to 45%. By contrast, the content of CaO changes slightly. This was due to the chemical reaction between [Mg] and Al_2O_3 inclusions formed in Al deoxidation. With the rise of reaction time from 60 to 90 min, the content of Al_2O_3 in inclusions increased from 40% to 45%, while the content of MgO decreased from 45% to 20%. By comparison, the content of Cao increased from 15% to 30%. This reflected the transformation of spinel type inclusions into calcium magnesia aluminate. With the increase of reaction time from 90 to 180 min, contents of MgO in inclusions continued to decrease from about 20% to 15%, while the content of Al_2O_3 increased from 45% to nearly 50%. Whereas, obvious changes in the content of CaO were not witnessed. In Fig. 5.6b, the average composition of inclusions in the ternary system at different reaction times were presented and compared. It can be seen that after 90 min' reaction, the average composition of inclusions had entered the region of melting point lower than 1773 K.

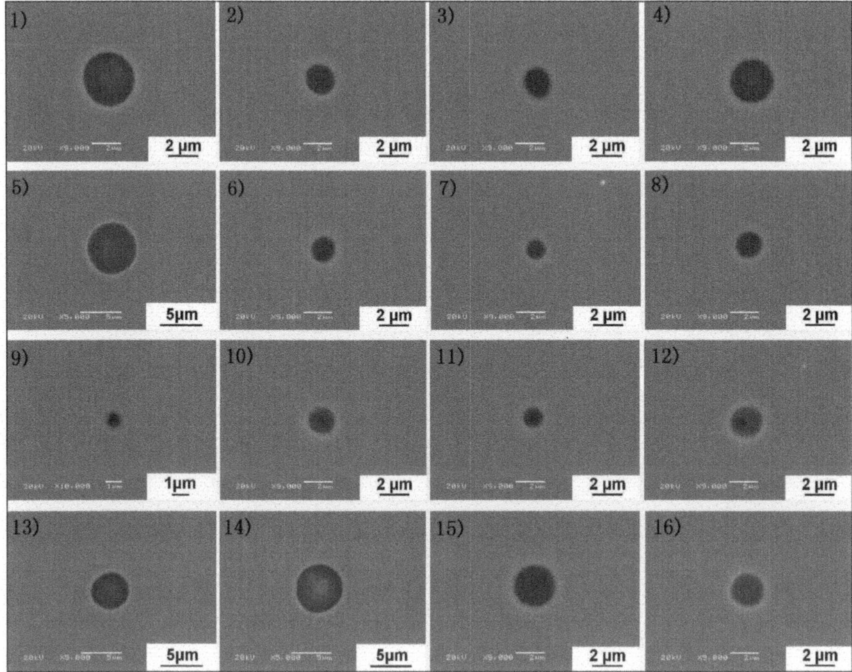

Fig. 5.3 Typical inclusion in molten steel reacted with slag for 60 min

In particular, it was noticed that although the reaction between molten steel and slag can approach equilibrium in about 90 min, variations in the chemical composition of inclusions still continued. It meant that chemical reactions between molten steel and inclusions did not stop after 90 min' reaction but was still going on. Therefore, it was necessary to study the inclusions in the transformation process.

5.1.3 Characterization on the Features of Inclusions at Different Reaction Time

In order to study the transformation mechanism of calcium magnesia aluminate inclusions in steel, features of the typical inclusions at different slag-steel reaction time were analyzed in details by SEM–EDS method with elment mappings and lining scannings.

The EDS analysis showed that, according to the chemical compositions, inclusions in steel were mainly composed of following system: (1) MgO-based inclusions; (2) $MgO{\cdot}Al_2O_3$ spinel inclusions; (3) CaO–MgO–Al_2O_3 inclusions. Sinel inclusions were in angular shape while the MgO-based inclusions were often in the form of periclase in steel, as shown in Fig. 5.7. The EDS spectrum of these inclusions were

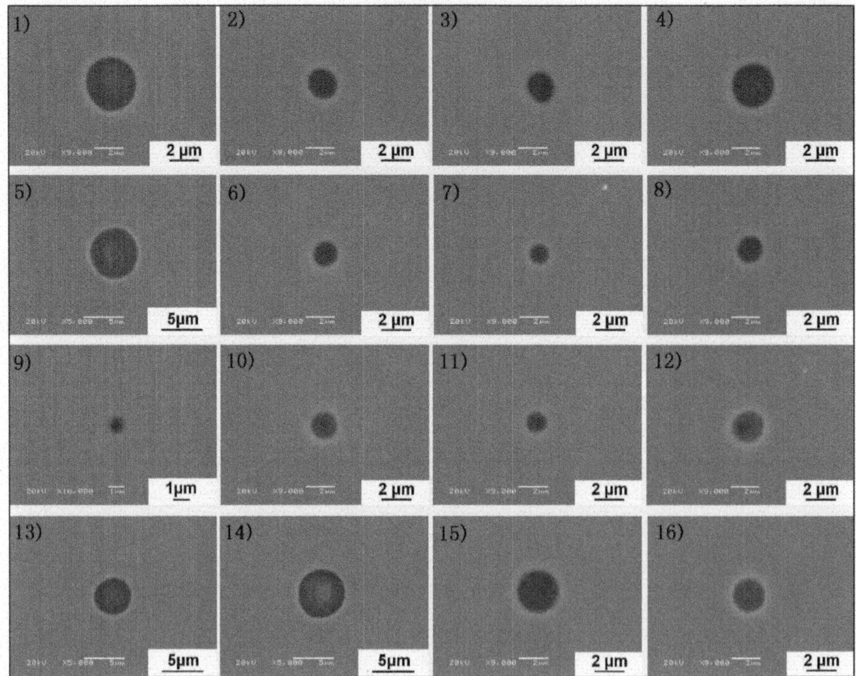

Fig. 5.4 Typical inclusion in molten steel reacted with slag for 90 min

shown in Table 5.8. Because of high melting point and poor deformability, these two kinds of inclusions would be very harmful to the fatigue resistance property of steel.

In addition, some $CaO–MgO–Al_2O_3$ inclusions were also formed in the steel after 30 min of slag-steel reaction. Such inclusions were characterized by spinel in the center and contained a certain amount of Ca. Particularly, such inclusion were in spherical shape, as shown in Fig. 5.8.

It can be seen from the energy spectrum analysis results of inclusions in Table 5.9 that the atomic percentages of Mg and Al in the inclusions observed in Fig. 5.8 are very close to 1:2, which was very consistent with the chemical composition of spinel.

In order to have a more accurate information on the element distribution in the whole inclusions, the element mappings of inclusions were carried out, as shown in Fig. 5.9. It can be seen that after the slag-steel reactions at 1873 K for 30 min, core of inclusion were mainly composed of spinel, while the surface was with higher content of CaO.

During the SEM–EDS analysis of inclusions, calcium magnesia aluminate inclusions with high MgO content spinel as the core surrounded by a high CaO content surface layer were observed. It should be pointed out that the MgO content in the core of this kind of inclusion was very high, which was distinctive from spinel in

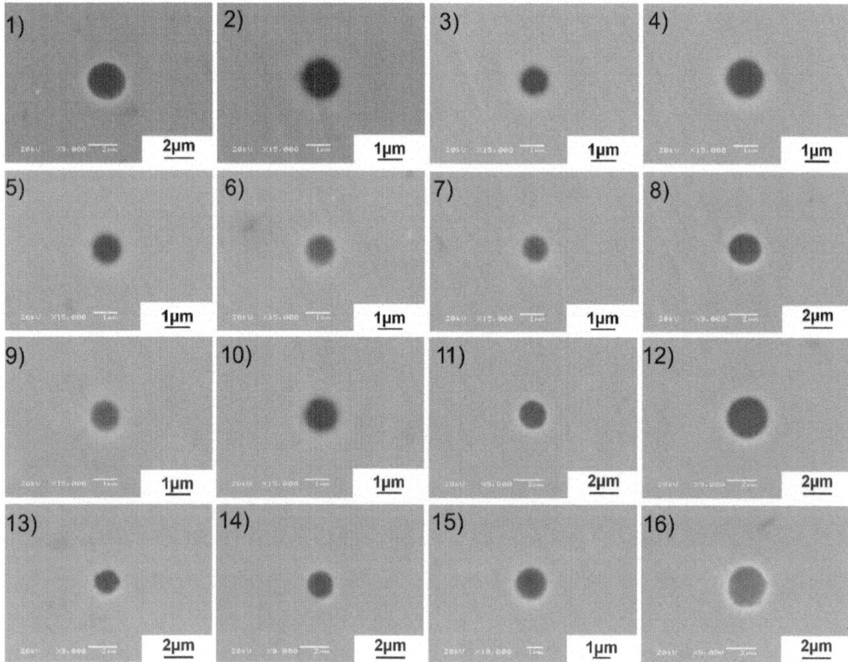

Fig. 5.5 Typical inclusion in molten steel reacted with slag for 180 min

Table 5.4 Chemical composition of inclusions in molten steel/slag reaction for 30 min, mass%

No.	Al_2O_3	SiO_2	CaO	MgO	Sum	No.	Al_2O_3	SiO_2	CaO	MgO	总和
1	65.9	0.0	7.4	26.7	100	9	51.2	3.5	34.2	11.2	100
2	0.0	0.0	0.0	100.0	100	10	0.0	0.0	0.0	100.0	100
3	0.0	2.2	0.0	97.8	100	11	51.5	6.9	30.6	11.0	100
4	50.8	3.0	40.9	5.3	100	12	70.8	0.0	3.4	25.8	100
5	72.2	0.0	3.4	24.4	100	13	67.8	0.0	0.0	32.2	100
6	6.4	3.3	5.7	84.6	100	14	66.9	0.0	0.0	33.1	100
7	56.8	0.0	12.8	30.4	100	15	52.1	4.4	33.2	10.3	100
8	64.8	3.0	3.6	28.6	100	16	70.8	0.0	2.1	27.1	100

chemical composition. Morphology and energy spectrum analysis result of this type of inclusions were shown in Fig. 5.10.

In order to have an overall recognition on the element distribution in inclusions, SEM line scannings of inclusions were analyzed. It was found that content of Mg in the core of inclusion was very high, while Al and Ca mainly distributed in the outer surface layer, as shown in Fig. 5.11. Hence, it can be concluded that core of

Table 5.5 Chemical composition of inclusions in steel reacted with slag for 60 min (mass%)

No.	Al_2O_3	SiO_2	CaO	MgO	Sum	No.	Al_2O_3	SiO_2	CaO	MgO	Sum
1	64.8	1.7	7.0	26.5	100	9	33.9	1.4	20.5	44.2	100
2	40.7	7.2	18.8	33.3	100	10	2.0	0.8	6.4	90.8	100
3	3.5	1.9	11.2	83.4	100	11	32.4	3.4	30.4	33.8	100
4	42.7	2.7	44.0	10.6	100	12	56.0	3.6	32.9	7.5	100
5	30.0	2.4	16.5	51.1	100	13	22.4	1.4	17.0	59.2	100
6	55.1	4.4	24.6	15.9	100	14	53.8	1.1	10.6	34.5	100
7	70.9	0.0	2.1	27.0	100	15	0.0	1.0	1.5	97.5	100
8	67.8	0.0	4.6	27.6	100	16	52.3	3.1	35.1	9.5	100

Table 5.6 Chemical composition of inclusions in steel reacted with slag for 90 min (mass%)

No.	Al_2O_3	SiO_2	CaO	MgO	Sum	No.	Al_2O_3	SiO_2	CaO	MgO	Sum
1	56.1	4.1	30.5	9.3	100	9	49.6	7.7	16.9	25.8	100
2	35.7	4.3	33.2	26.8	100	10	53.9	6.0	18.8	21.3	100
3	47.8	3.2	43.1	5.9	100	11	47.5	4.1	43.0	5.4	100
4	47.1	5.2	24.8	22.9	100	12	44.8	3.5	45.9	5.8	100
5	46.8	6.6	20.3	26.3	100	13	47.0	5.2	23.7	24.1	100
6	54.6	5.5	31.9	8.0	100	14	51.1	7.5	21.5	19.9	100
7	37.7	5.3	23.6	33.4	100	15	41.4	3.9	43.5	11.2	100
8	57.1	6.2	23.0	13.7	100	16	47.2	5.3	39.6	7.9	100

Table 5.7 Chemical composition of inclusions in steel reacted with slag for 180 min (mass%)

No.	Al_2O_3	MgO	CaO	SiO_2	Sum	No.	Al_2O_3	MgO	CaO	SiO_2	Sum
1	52.7	11.5	31.3	4.5	100	9	40.6	38.1	16.0	5.4	100
2	37.8	29.7	29.3	3.3	100	10	46.4	34.8	16.5	2.3	100
3	32.0	44.3	20.5	3.2	100	11	46.0	5.3	44.6	4.1	100
4	56.0	12.0	27.2	4.8	100	12	50.8	16.8	27.4	50.8	100
5	44.5	28.9	22.3	4.3	100	13	41.4	34.5	19.5	4.6	100
6	44.7	36.6	12.9	5.7	100	14	59.3	8.7	27.8	4.2	100
7	50.9	28.1	15.6	5.4	100	15	41.1	31.5	22.3	5.2	100
8	53.2	8.2	34.0	4.7	100	16	48.9	23.1	23.5	4.4	100

the inclusion was mainly composed of MgO while the outer layer was composed of Al_2O_3 and CaO.

Table 5.10 showed the analysis results of contained elements in the inclusion in Fig. 5.7 by the EDS method. As can be seen, the atomic percentage of Mg and Al in the core was about 1:1, which was different from that of Mg and Al in spinel.

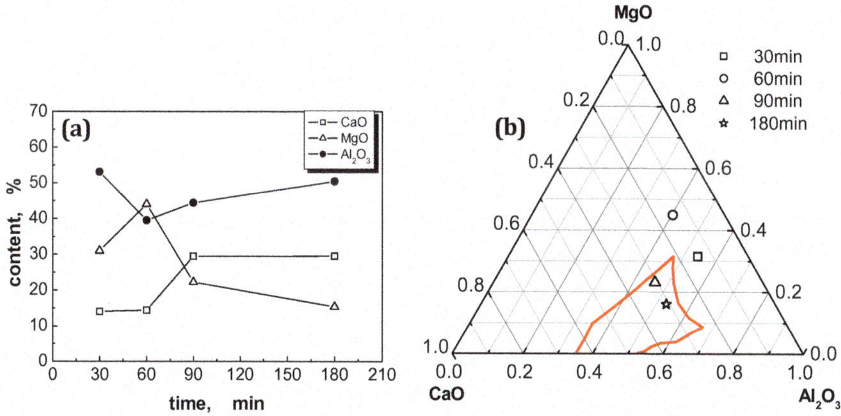

Fig. 5.6 Variation in average composition of inclusions with reaction time, reprinted from a previous paper of present authors on ISIJ International, 2010, vol. 50, pp. 96–104, with permission from ISIJ

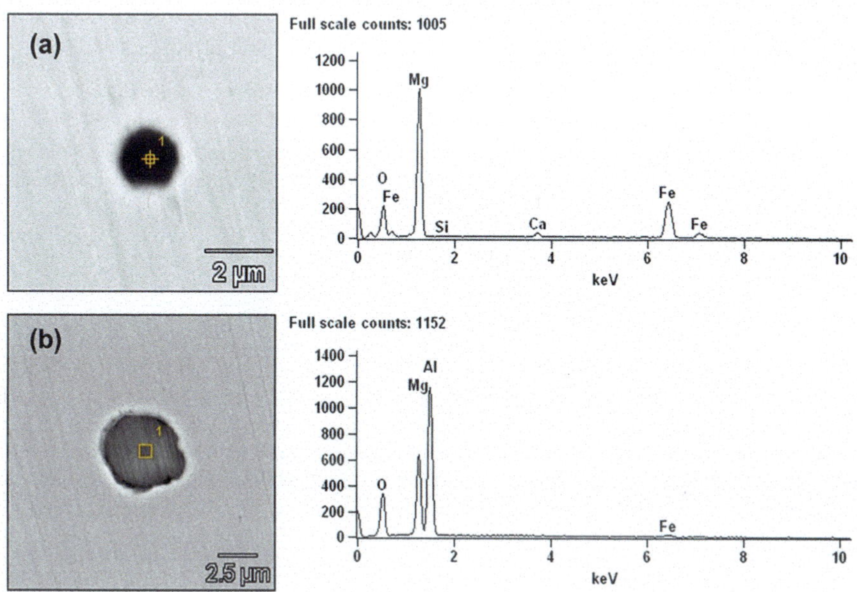

Fig. 5.7 **a** MgO inclusions; **b** $MgO{\cdot}Al_2O_3$ inclusions

Table 5.8 Inclusion spectrum analysis results of MgO and spinel, atom%

	O	Mg	Al	Si	Ca	Fe
(a)	38.81	44.22	–	0.28	0.86	15.84
(b)	54.87	14.25	30.21	–	–	0.68

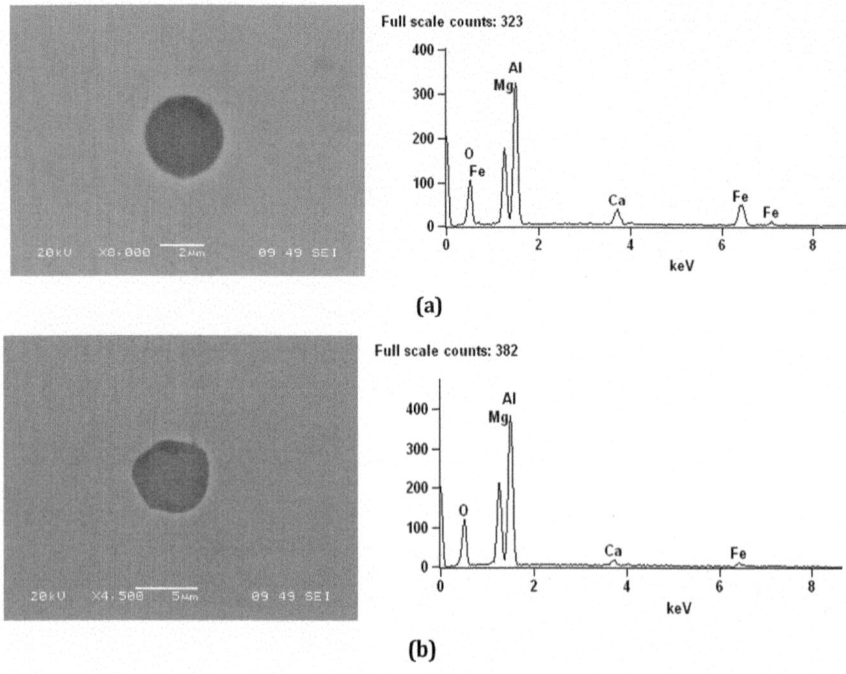

Fig. 5.8 Typical calcium magnesia aluminate with spinel as the core

Table 5.9 EDS analysis results of inclusion, atom%

	O	Mg	Al	Ca	Fe
(a)	50.25	13.64	26.40	2.71	7.00
(b)	53.99	14.45	29.33	0.95	1.27

Therefore, it can be judged that the core of this kind of inclusion was not pure spinel but was a type of MgO–Al_2O_3 compound with high MgO content.

Therefore, according to chemical composition of the core, CaO–MgO–Al_2O_3 inclusions in steel can be further divided into two categories: ① the first type was with high Al_2O_3 content MgO–Al_2O_3 as the core; ② the second type was featured by high MgO content MgO–Al_2O_3 as the core.

By comparison, natures of inclusions like composition and type were not obvious changed after 60 min. Inclusions were still of three types: (1) MgO-based inclusions; (2) Al_2O_3–MgO spinel inclusions; (3) calcium magnesia aluminate inclusions. During the SEM–EDS analysis, singular MgO-based inclusions and calcium magnesia aluminate inclusions with Al_2O_3–MgO cores were observed, as shown in Figs. 5.12 and 5.13, respectively.

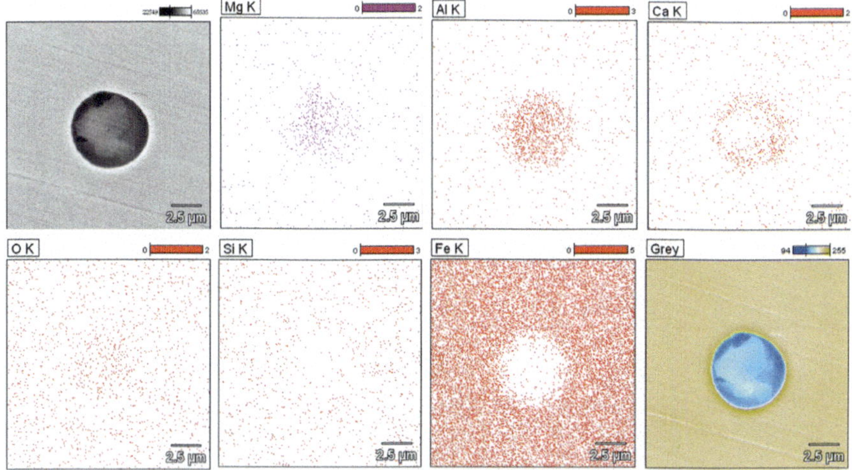

Fig. 5.9 SEM mapping of inclusions with spinel as the core, reprinted from a previous paper of present authors on ISIJ International, 2010, vol. 50, pp. 96–104, with permission from ISIJ

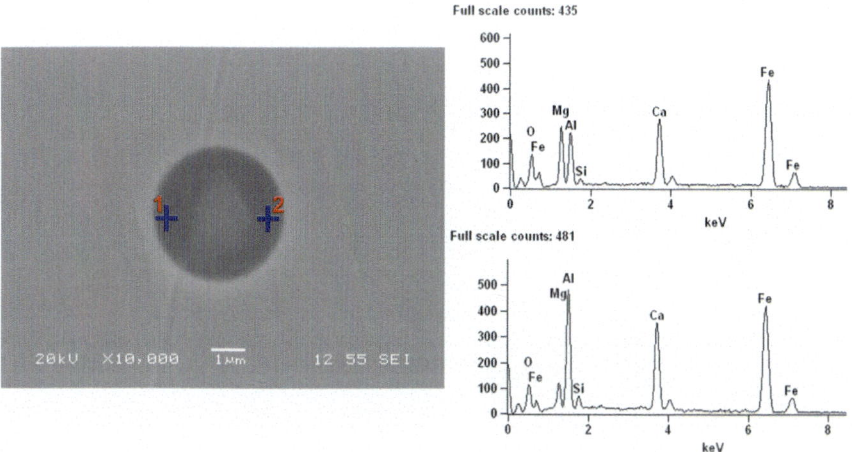

Fig. 5.10 Calcium magnesia aluminate with high MgO content spinel core

It can be found from the above results that some MgO-based inclusions in steel already contain a certain amount of Al content and Ca content after 60 min' of reaction, compositions of such typicl inclusions were shown in Table 5.11.

SEM-mapping of inclusion in Fig. 5.14 showed that core of the inclusion was MgO-based while the surface of the inclusions was wrapped by a layer of CaO–Al_2O_3. There was a layer with high content of Al and Ca at the edge of the inclusion,

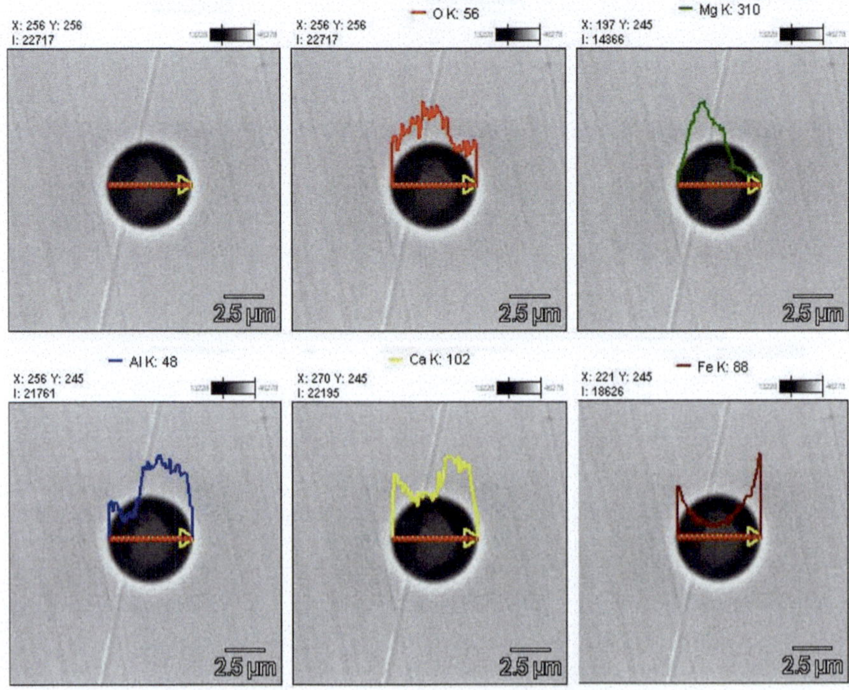

Fig. 5.11 Line scanning of calcium magnesia aluminate inclusions

Table 5.10 EDS analysis result of inclusions (atom%)

	O	Mg	Al	Si	Ca	Fe
1	35.78	13.42	10.24	0.94	9.61	30.01
2	36.11	4.77	18.82	1.52	12.11	26.67

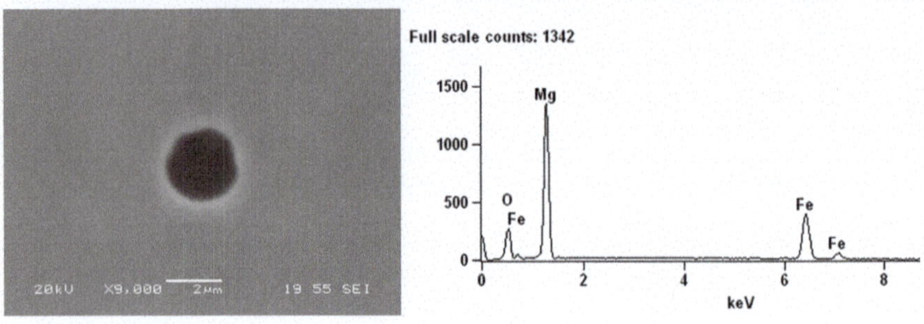

Fig. 5.12 MgO-based inclusion

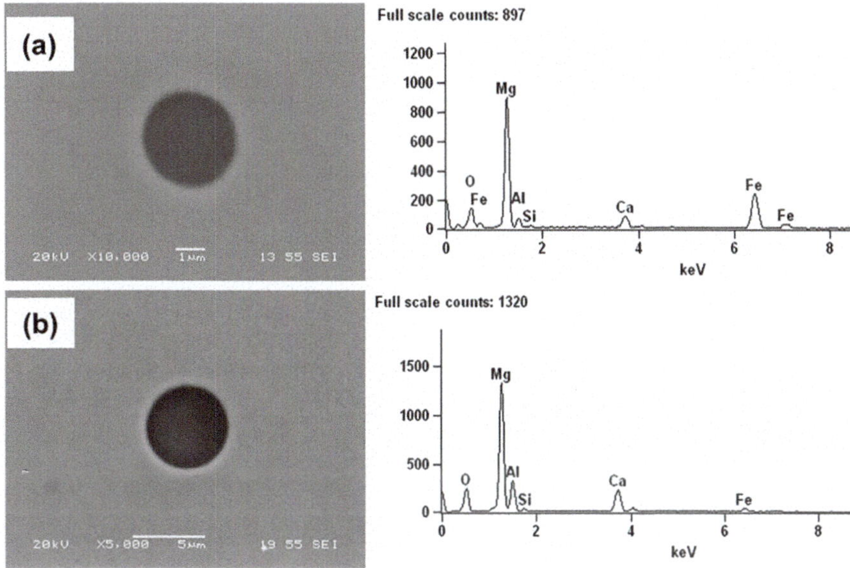

Fig. 5.13 MgO-based inclusions under tranferring

Table 5.11 Compositions of tranferring MgO-based inclusions at 60 min (atom%)

	O	Mg	Al	Si	Ca	Fe
a	41.78	42.76	0.77	0.25	2.14	12.30
b	46.03	45.82	1.60	0.26	0.12	1.80
c	50.57	31.63	9.37	0.50	6.50	1.43

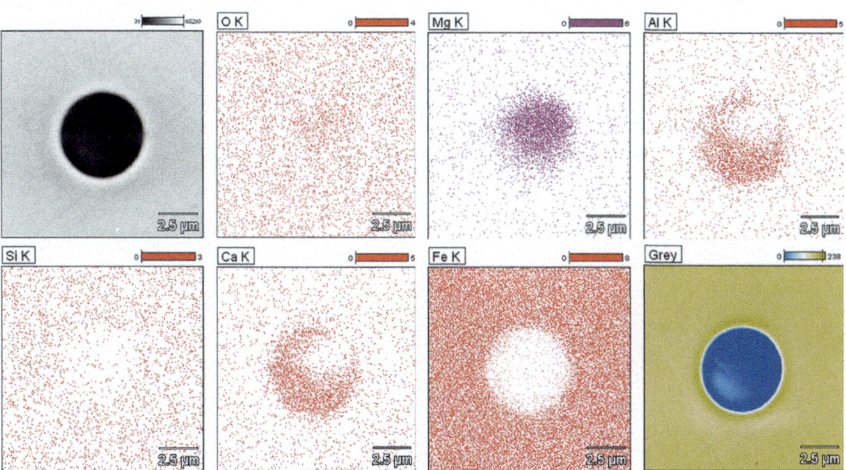

Fig. 5.14 SEM mapping of typical inclusions with MgO-based cores, reprinted from a previous paper of present authors on ISIJ International, 2010, vol. 50, pp. 96–104, with permission from ISIJ

while the core of inclusion was featured by very high content of Mg. The enriched layer of Al and Ca overlapped each other while complementary with that of Mg.

Figures 5.15 and 5.16 showed the EDS results and SEM-mappings of Al_2O_3–MgO type spinel experiencing the transformation.

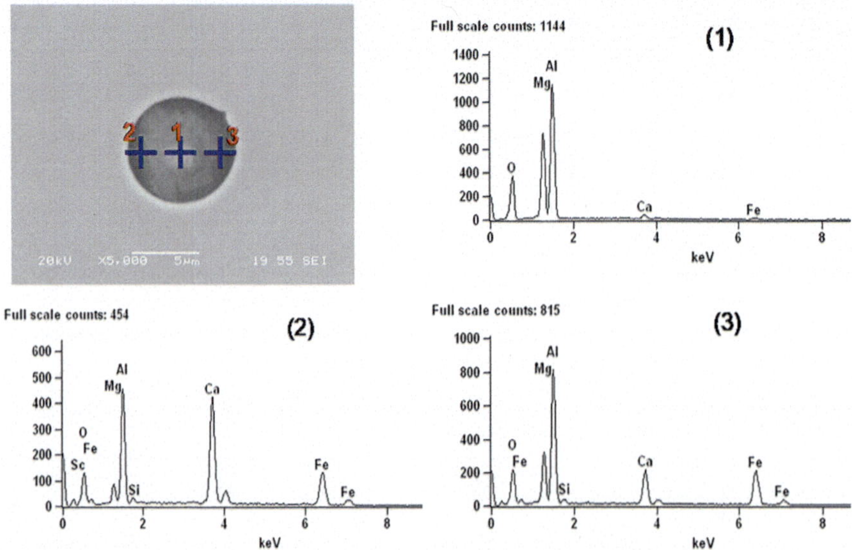

Fig. 5.15 SEM–EDS analysi results of Al_2O_3–MgO spinel experiencing transformation

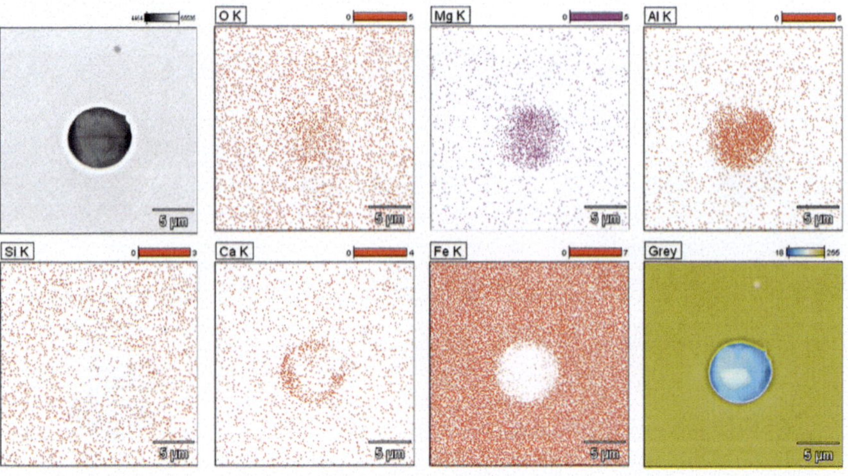

Fig. 5.16 SEM mapping of Al_2O_3–MgO spinel experiencing evolution, reprinted from a previous paper of present authors on ISIJ International, 2010, vol. 50, pp. 96–104, with permission from ISIJ

Table 5.12 Atomic percentage of elements in inclusion (atom%)

	O	Mg	Al	Si	Ca	Fe
b-1	55.81	15.34	27.15	0	1.04	0.66
b-2	52.89	2.98	17.96	0.87	15.61	9.48
b-3	47.10	10.05	24.70	0.89	6.23	11.04

It can be seen that after 60 min' reaction, spinel inclusions in steel were being transformed into calcium magnesia aluminate inclusions. As a result, enrichement of Ca and Mg in the inclusions supplemented to each other without overlapping. Atomic percentages of included elments in the inclusion were shown in Table 5.12. From the data, it can be seen that the inclusion was with spinel as the core surrounded by an outer surface layer of CaO–Al_2O_3.

Whereas, after 60 min' reaction, there were still single spinel inclusions in steel, as typical shown in Fig. 5.17.

After 90 min' reaction, composition of inclusions in steel obviously changed and distinctive from inclusions at 60 min: (1) all the observed inclusions were composed of calcium magnesia aluminate; (2) singular spinel and single MgO-based inclusions cannot be observed in steel. Figure 5.18 showed the EDS analysis results and SEM mappings of the calcium magnesia aluminate inclusions in steel. It can be seen from the figure that the content of MgO in the center of inclusion was very high while the contents of Al_2O_3 and CaO were much higher in the ourter part of the inclusion. The enrichment areas of Al and Ca overlapped each other, but supplementary to the enrichment of Mg.

Figure 5.19 indicated the SEM mappings of the observed two types of calcium magnesia aluminate at 90 min. It can be clearly seen from the figures that CaO contents in inclusions were relatively higher than that at 60 min. Moreover, content of MgO in the center of inclusion was very high, while the outer layer of inclusion was wrapped by a layer with high contents of CaO and Al_2O_3.

After 180 min of slag-steel reaction, all the observed inclusions in steel were composed of calcium magnesia aluminate. And there were neither singular MgO nor single spinel inclusions as well as high MgO content inclusions. Figure 5.20 showed

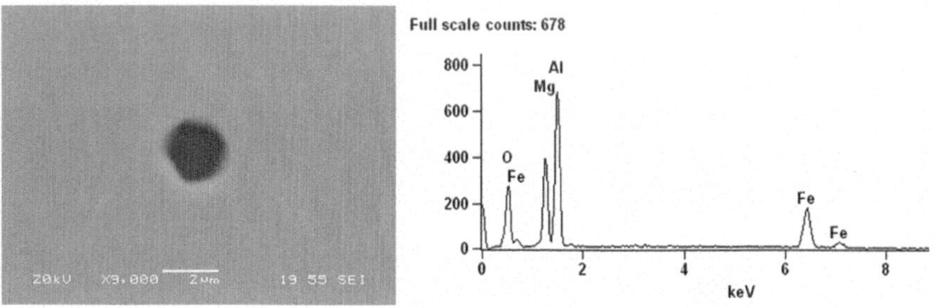

Fig. 5.17 Typical Al_2O_3–MgO type spinel inclusions in steel at 60 min

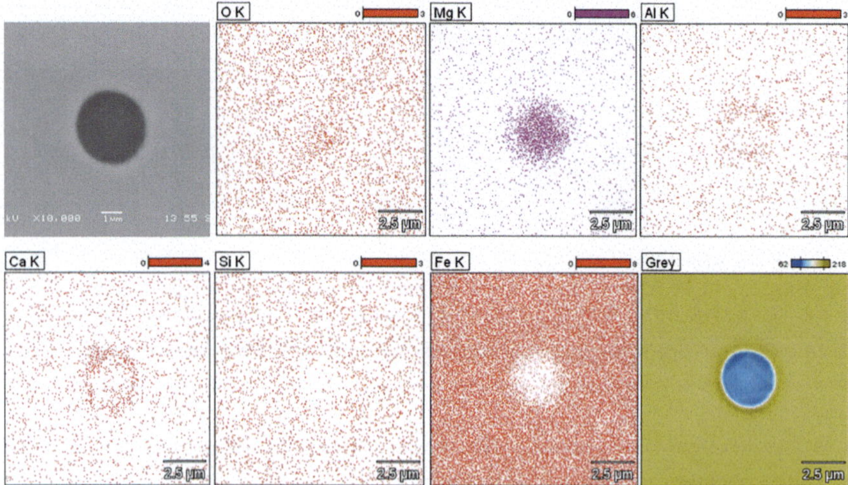

Fig. 5.18 SEM mapping of MgO-based inclusion evolving into calcium magnesia aluminate

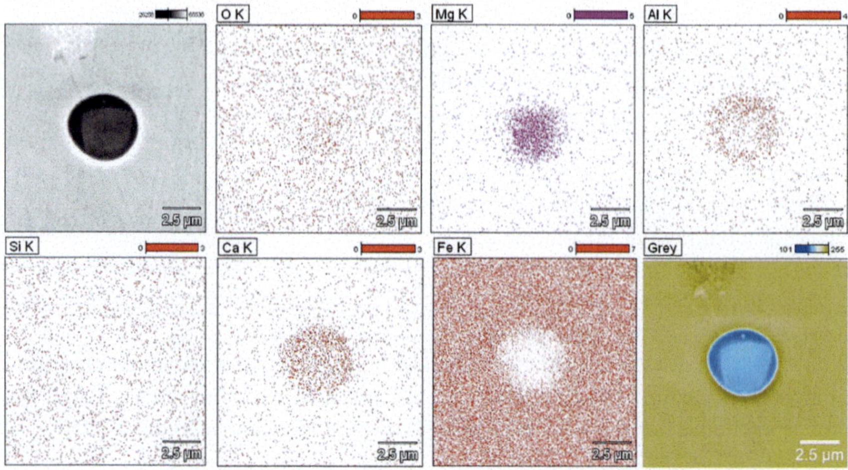

Fig. 5.19 SEM mappings of spinel evolving into calcium magnesia aluminate

the SEM EDS results of typical inclusions in steel at 180 min. Atomic percentages of included elements obtained the EDS were shown in Table 5.13. It was found from the SEM EDS analysis that atomic percentages of Mg and Al in the inclusion (a) were approximately about 1:2, with obviously increased content of CaO, compared to the result of 90 min. In inclusion (b), the content of MgO in the core was very high, while the contents of CaO and Al_2O_3 at the surface were very high.

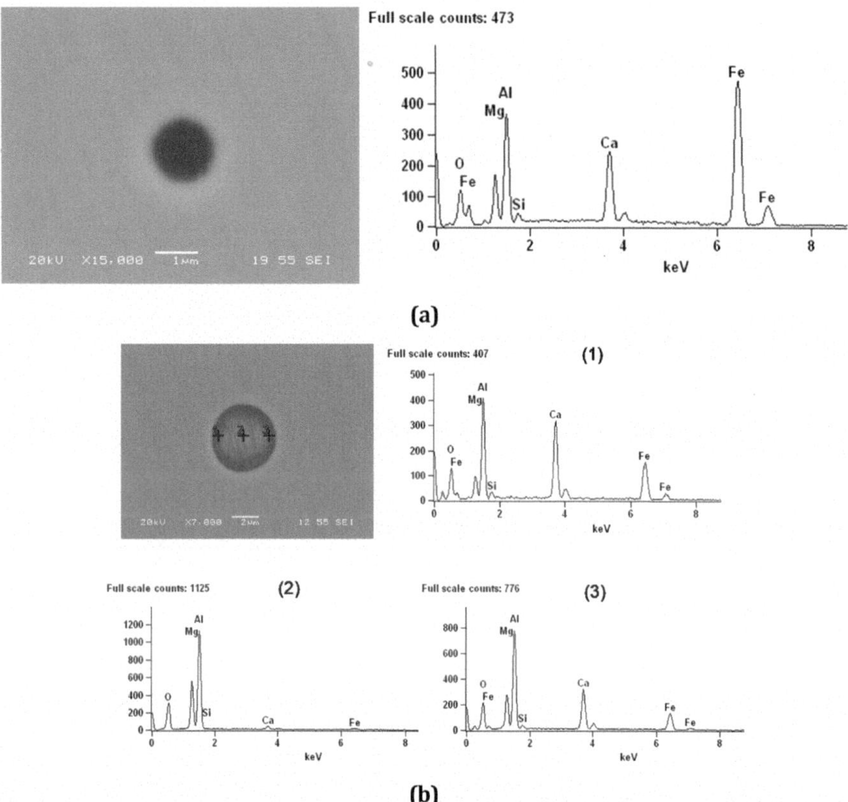

Fig. 5.20 SEM EDS results of calcium magnesia aluminate inclusions in steel

Table 5.13 Inclusion spectrum analysis results, atom%

	O	Mg	Al	Si	Ca	Fe
a	32.80	8.05	16.58	1.03	8.95	32.60
b-1	50.56	4.74	17.89	1.09	13.35	12.37
b-2	55.14	13.14	29.88	0.21	0.98	0.66
b-3	53.62	8.10	22.06	0.71	8.58	6.92

Figure 5.21 was the SEM line scanning of an inclusion. It can be seen that core of the inclusion was enriched with $MgO–Al_2O_3$, while the outer surface layer was composed of CaO and Al_2O_3, which were consistent with EDS analysis.

Based on above obtained results, evolutions of the inclusions in steel with the rise of slag-steel reaction time can be known. The results indicated that with reaction time increased from 30 min to 6o minutes, 90 min and 180 min, inclusions were changed from Al_2O_3 to spinel and finally low melting point calcium magnesia aluminate.

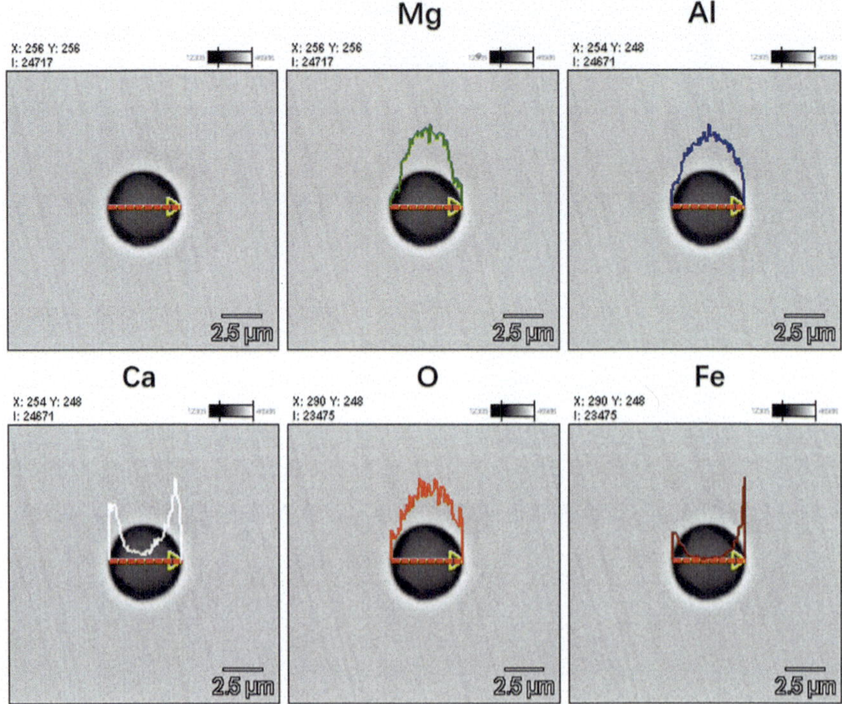

Fig. 5.21 SEM line scanning of inclusion after 180 min' reaction

Particularly, the calcium magnesium aluminate inclusions were featured by spinel as the core, wrapped by an outer surface layer of CaO–Al_2O_3.

And the spinel cores can be also divided into two kinds. One was featured by high Al_2O_3 content, and marked as Al_2O_3–MgO in the context. While the other type of spinel core was featured by high MgO content, therefore, written as MgO–Al_2O_3. By comparison, the former type of core occupied a much larger fraction while latter was only observed in a small part of inclusions.

5.2 Descriptions on the Evolutions of Inclusions

According to the obtained results, it can be known that calcium magnesia aluminate inclusions in steel were transferred spinel inclusions. There were some calcium magnesia aluminate inclusions with higher MgO content, however, they occupied a small fraction. Hence, emphasis was put to the transformation of the former type of inclusions.

Based on SEM–EDS and SEM mapping of inclusions at varied reaction time, a macro recognition on the changes of inclusions can be plotted. However, at the high temperatue of steelmaking, changes of inclusions in the liquid steel was a black box. Hence, it was necessary to infer and explore the transformation mechanism of inclusions at high temperature, based on the features of inclusions in steel at the room temperature.

In this section, the transformation mechanism of calcium magnesia aluminate inclusions with lower melting point in spring steel would be further explored and discussed from the perspective of slag-steel reaction equilibrium.

Before the explanations on transformation process of the calcium magnesia aluminate inclusions in the experiment, several aggreements reached in published literatures were summarized as following:

(1) During slag-steel reaction experiments, interactions among slag, molten steel and inclusions can be summarized briefly. That was: complete equilibrium among slag, molten steel and inclusions were very difficult to establish but the local equilibrium among them can absolutely be achieved. As the slag directly contacts with molten steel, compositions of slag would directly affect the contents of [Ca], [Mg], [Al] and other strong deoxidizing elements as well as the contents of [O] in molten steel through slag-steel chemical reactions While the contents of [Ca], [Mg], [al], [O] in liquid steel would directly determine the composition of inclusions in steel by the chemical reactions between steel and inclusions.

(2) With the use of strong Al deoxidation, [Al] in steel can react with the slag and reduce MgO and Cao in the slag, resulting in the increase of [Mg] and [Ca] contents in the steel.

(3) Al_2O_3 inclusions in steel formed in the Al deoxidation would react with [Mg] to proudce spinel inclusions. As a result, spinel inclusions would be largely produced in the steel.

(4) When there is a certain amount of [Ca] in liquid steel, spinel inclusions would be unstable and be changed into calcium magnesia aluminate inclusions. During the reaction, [Mg] in the inclusion would be replaced by [Ca] in molten steel. When the slag contains a certain content of SiO_2, SiO_2 would consume [Ca] thus stabilizing the existence of spinel.

According to the interaction mechanism among slag, liquid steel and inclusion, chemical reaction among slag-liquid steel-inclusion in the present study was illustrated as Fig. 5.22. At the very beginning of slag-steel reaction, because of Al deoxidation, a large number of Al_2O_3 inclusion particles would be generated in steel due to Al deoxidation. Afterwards, with the extension of reaction time, larger Al_2O_3 inclusions would be floated out into the slag for removal. However, a large number of small particles of Al_2O_3 inclusions would still remain in the molten steel. Subsequently, due to the reaction between slag and molten steel, CaO and MgO in slag would be reduced by [Al] in molten steel, providing [Ca] and [Mg] into molten steel, as shown in Fig. 5.23.

Later, the formed Al_2O_3 inclusions would react with [Mg] in the steel. As a result, a large number of Al_2O_3–MgO spinel inclusions would be generated in liquid steel, as shown in Fig. 5.24.

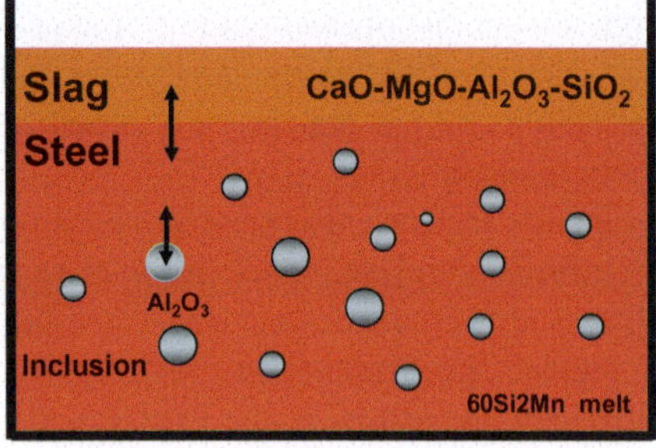

Fig. 5.22 Al deoxidation of molten steel

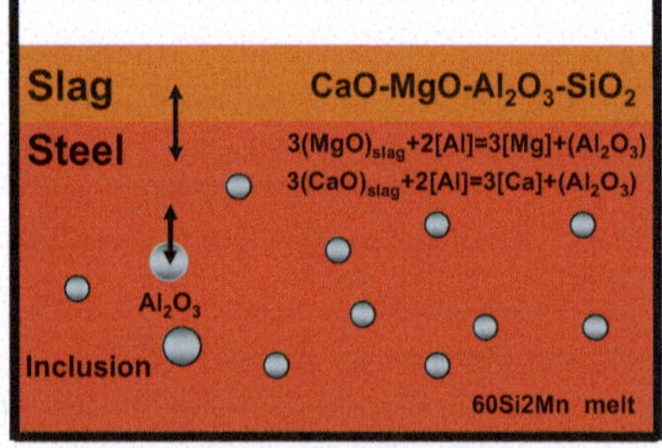

Fig. 5.23 Slag-steel reaction

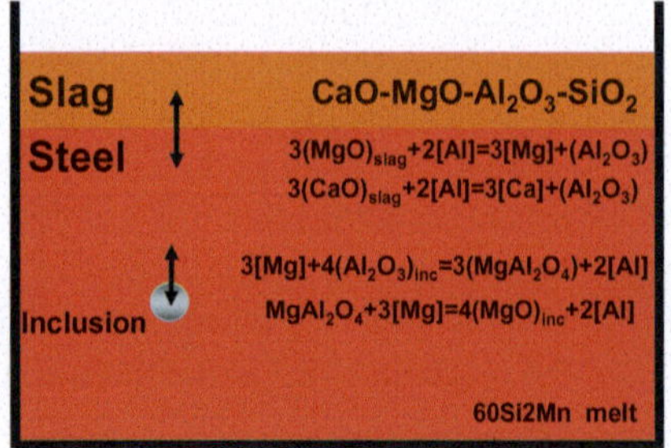

Fig. 5.24 Formation of Al_2O_3–MgO spinel

Due to the presence of a certain [Ca] content in the liquid steel, the Al_2O_3–MgO spinel inclusions would react with [Ca] and were gradually modified into calcium magnesia aluminate inclusions, as shown in Fig. 5.25. Because Mg in the inclusion was replaced by [Ca] during the steel-inclusion the reaction, enrichments of Ca and Mg in inclusions supplemented to each other.

Based on the above discussions, the following chemical reaction equation can be deduced to brifely summarize the formation and transformation of calcium magnesia aluminate inclusions in steel:

$$x[\mathrm{Ca}] + (x + y + 3z)/3(\mathrm{Al_2O_3}) + y[\mathrm{Mg}] = (x\mathrm{CaO} \cdot y\mathrm{MgO} \cdot z\mathrm{Al_2O_3}) + (2x + 2y - 3z)[\mathrm{Al}] \quad (5.1)$$

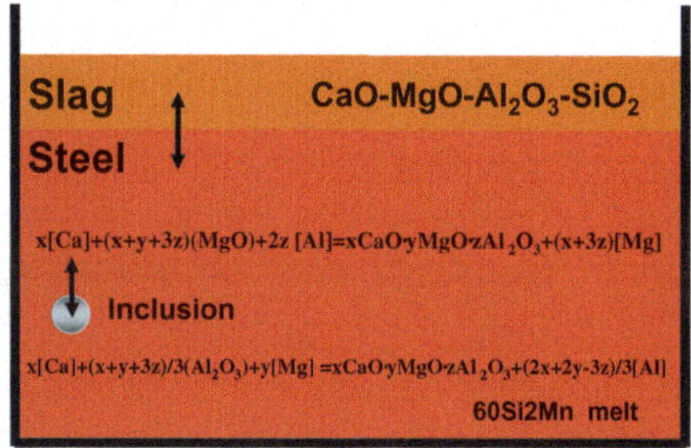

Fig. 5.25 Evolution of spinel to calcium magnesia aluminate

Chemical reaction Eq. (5.1) can be divided into two stages: (1) the first stage was the transformation of Al_2O_3 into Al_2O_3–MgO spinel; (2) the subsequent stage was the transformation of Al_2O_3–MgO spinel into calcium magnesia aluminate inclusions.

As a result, transformation of calcium magnesia aluminoate inclusions in steel can be summarized as following:

(1) At the beginning of slag-steel reaction, a large number of deoxidation products of Al_2O_3 were formed in the steel;

(2) With the increase of reaction time, due to the errosion of slag on MgO crucible, MgO dissolved into slag phase, witnessing increased MgO in slag phase. Then, chemical reaction would take place between molten steel and slag, viz., [Al] would reduce MgO and CaO in slag to supply [Mg] and [Ca] into molten steel. Hence,the contents of [Mg] and [Ca] in molten steel increased.

(3) Al_2O_3–MgO spinel inclusions would be formed in molten steel due to the chemical reaction between Al_2O_3 inclusions and [Mg] in molten steel. Although there was a certain content of [Ca] in molten steel at the initial stage of slag-steel reaction, the content [Ca] was limited. In addition, as previously mentioned, SiO_2 in slag would stabilize the formation of spinel inclusions to a certain extent Therefore, only part of the Al_2O_3–MgO spinel inclusions in steel were transformed into calcium magnesia aluminate at the early stage of expeirments.

(4) With the increase of reaction time, due to the presence of increased amount of [Ca] in steel, the spinel inclusions would be in an unstable state and gradually transformed into calcium magnesia alumiante inclusions.

(5) During the transformation from spinel to calcium magnesia alumiante inclusions, Mg in the inclusion would be replaced and reduced by [Ca]. Therefore, the spatial distribution of Mg and Ca in inclusions would be complementary.

5.3 Evolution Mechanisms of Inclusions

Although the formation and transformation process of calcium magnesia aluminate inclusions in steel were described in details. However, there were few discussions on the transformation mechanism and kinetics of spinel inclusions during the slag-steel reactions. For this reason, this paper makes a preliminary and qualitative discussion on the transformation kinetics of inclusions.

5.3.1 Concentration Gradients of Elements in Inclusions

During the line scanning analysis of inclusions under SEM–EDS, it was found that the contents of Ca, Mg and Al in inclusions have obvious concentration gradients along the scanning rountine line. Figure 5.26 showed line scan results of the inclusions of a Al_2O_3–MgO spinel experiencing transformation. It can be seen that the variation tendency of Al and Ca in inclusions was opposite to that of Mg content. In the area where Mg content decreased, Al and Ca contents increased obviously.

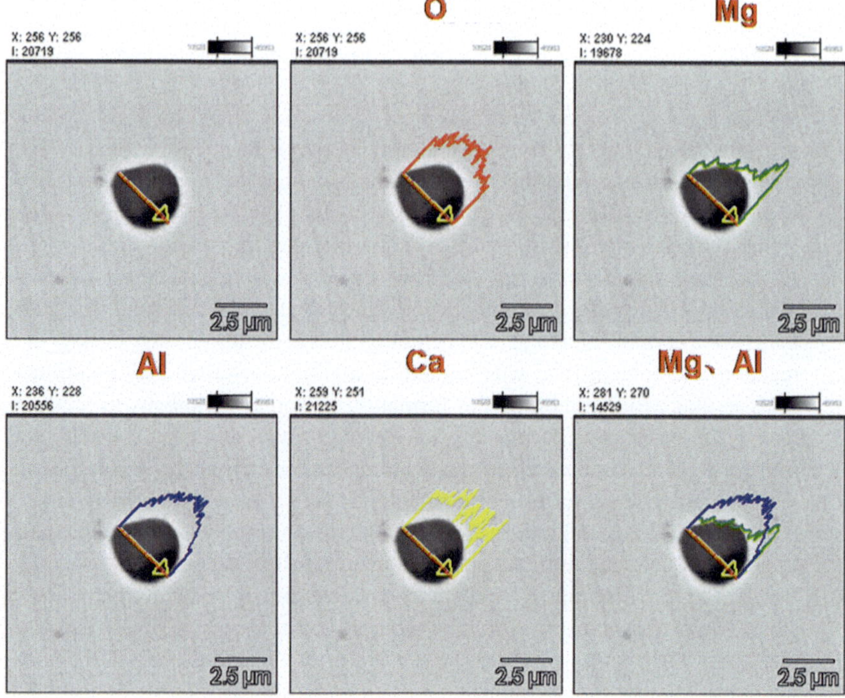

Fig. 5.26 Al_2O_3–MgO spinel experiencing transformation, e, reprinted from a previous paper of present authors on ISIJ International, 2010, vol. 50, pp. 96–104, with permission from ISIJ

5.3.2 Model Estabishment on the Evolution of Inclusions

Ye et al. had discussed the transformation kinetics of calcium aluminate inclusions during calcium treatment. The results showed that with the increase of reaction time, the content of CaO in the inclusions would continuously increase and the compositions of inclusions would be homogenized by diffusion. As a result, the inclusions would be finally transformed from alumina into liquid calcium aluminate. However, this model cannot explain how CaO entered into the inclusions and how it diffused and homogenized in the solid inclusions. As it known, in the industrial practice, characteristics of ionic bond/chemical bond in solid compounds determined that direct diffusion of CaO in solid inclusions would be very difficult [1].

Parra and Allibert studied the physicochemical reaction process of Li_2O–Al_2O_3–SiO_2 slag with Cu–Ni–Al alloy, discussing in details the diffusion between the slag and alloy during the reaction. The results revealed the existence of diffusion phenomenon and established the ion exchange reaction model between slag and alloy. The interface reaction between alloy and slag was explained based on the ion reactions, and the reaction mechanism can be schematically shown in Fig. 5.27 [2].

Park explained the transformation process and mechanism of Al_2O_3 inclusions into calcium aluminate during the calcium treatment of stainless steel. Combining the understandings on the microstructures of CaO–Al_2O_3 slag, he proposed the chemical reaction equation between [Ca] and Al_2O_3 [3]. Based on detailed and careful analysis, Park believed that the diffusion of Al in the inclusion was the limiting step because the diffusion coefficient of Al was several orders of magnitude smaller than that of Ca.

Due to special crystal structure, physical and chemical properties, spinel has been widely studied in the field of solid chemistry [4, 5]. The results showed that when the molar ratio of MgO to Al_2O_3 was 1:1 and the reaction temperature was 1773 K for a certain time, the MgO–Al_2O_3 spinel compound layer can be formed between the MgO and Al_2O_3 solid layers. The well-known Wagner reaction mechanism shows that Mg^{2+} and Al^{3+} ions diffuse through the product layer, and then continue to react with the two reactants product and surface. The reaction process can be schematically shown in Fig. 5.28.

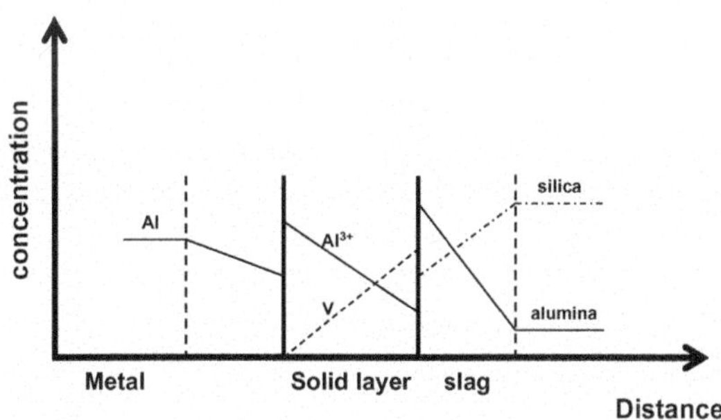

Fig. 5.27 Schematics of slag-metal reaction

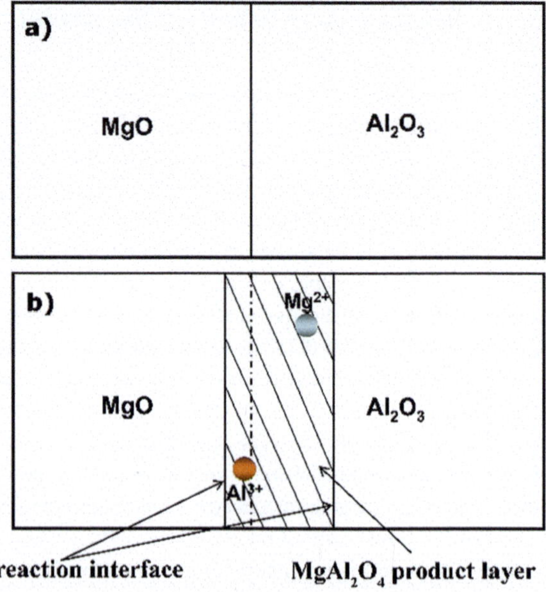

Fig. 5.28 Schematics on the formation of spinel by interface reaction between solids

In order to balance the charge, there would three Mg^{2+} would diffuses to the right interface when two Al^{3+} diffuses to the left interface. Ideally, the reaction at the two interfaces can be written as following. And during the reaction, the rate of reaction (5.2) would be three times of that in reaction (5.3).

(1) At $MgO/MgAl_2O_4$ boundary:

$$2Al^{3+} - 3Mg^{2+} + 4MgO \rightarrow MgAl_2O_4 \tag{5.2}$$

(2) At $MgAl_2O_4/Al_2O_3$ boundary:

$$3Mg^{2+} - 2Al^{3+} + 4Al_2O_3 \rightarrow 3MgAl_2O_4 \tag{5.3}$$

To simplify the discussion, all inclusions were considered as spherical balls. It was also assumed that the formation, transformation or growth of inclusions followed the spherically symmetric rule. The transformation process of Al_2O_3 to spinel can be shown in Fig. 5.29. Assume that there was a intermediate layer between the boundary layers at the inclusion side and at the steel side.

Steps involved in the chemical reactions between Al_2O_3 inclusion and molten steel can be described as following:

(1) Atoms of Al and Mg diffuse from liquid steel to the steel-inclusion interface.
(2) Because Mg is very active, Mg^{2+} can be produced by reaction (5.4) at the interface of steel-intermediate layer.

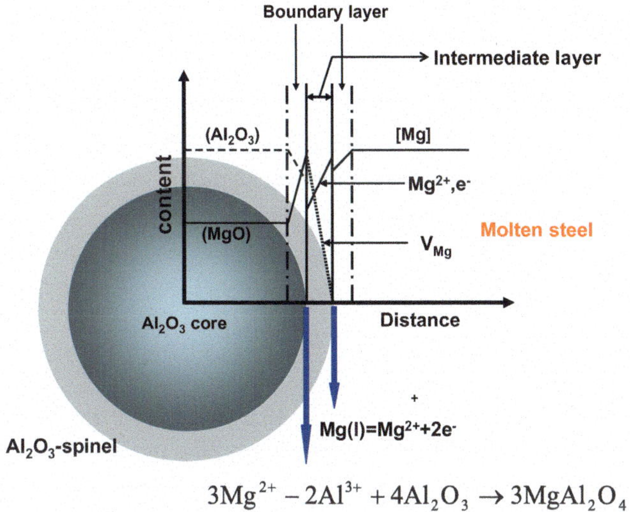

Fig. 5.29 Schematical description on the formation of Al_2O_3–MgO spinel, reprinted from a previous paper of present authors on ISIJ International, 2010, vol. 50, pp. 96–104, with permission from ISIJ

$$[\mathrm{Mg}] = \mathrm{Mg}^{2+} + 2\mathrm{e}^- \tag{5.4}$$

(3) Concentration gradients of Mg^{2+} and electrons would be formed at the interface of transition layer-steel. Then, they would pass through the transition layer by diffusion and gradually reached the interface of boundary layer-transition layer on the inclusion side. Meanwhile, vacancies with the same molar faction as cations would leave the transition layer by diffusion.

(4) Mg^{2+} diffused to the interface of inclusion-transition layer and then reacted with Al_2O_3 inclusions to produce Al_2O_3–MgO spinel inclusions.

$$3\mathrm{Mg}^{2+} + 4\mathrm{Al}_2\mathrm{O}_3 = 3\mathrm{MgAl}_2\mathrm{O}_4 + 2\mathrm{Al}^{3+} \tag{5.5}$$

(5) With the continuous proceeding of these reactions, the newly formed product would adhere to original inclusion particles to form a layer of the product, thus gradually forming a concentration gradient of the product along the radius direction of inclusion.

Zayan et al. showed that the diffusion of Mg ions in spinel inclusions was the rate limitation step in the reaction [6]. The formation rate of Al^{3+} at the interface of steel-transition layer and the diffusion rate of Al^{3+} in the transition layer were much higher than that of Mg^{2+} at the boundary between inclusions-transition layer.

According to the previous study, [7] reaction between spinel and [Ca] in steel can be shown as reaction (5.6). During the reaction, Mg would be replaced from the inclusion, and the inclusion would be transformed into calcium magnesia aluminate.

With the increase of reaction time, content of Ca in inclusions would increase.

$$2(MgO \cdot Al_2O_3) + [Ca] = (CaO{-}MgO{-}2Al_2O_3) + [Mg] \tag{5.6}$$

In present experiments, it was considered that the transformation mechanism of spinel inclusion to calcium magnesia alumiante inclusion included following steps.

(1) Diffusions of Ca atoms from liquid steel to liquid steel-inclusion interface.
(2) Because Ca is very active at high temperature, the reaction (5.7) would occur at the interface of liquid steel and intermediate layer.

$$Ca(l) = Ca^{2+} + 2e^- \tag{5.7}$$

(3) Concentration gradients of Ca^{2+} ions and electrons at the transition layer-liquid steel interface, would be formed, resulting in the diffusion of them to the interface of boundary layer-transition layer. Meanwhile, vacancies with the same molar fraction would leave the intermediate transition layer in the opposite direction through diffusion.
(4) With Ca^{2+} diffused to the inclusion-transition layer interface, [Ca] would react with spinel. During the reaction, Mg was gradually replaced from spinel and diffused away from inclusions through the boundary layer and intermediate transition layer, and finally into molten steel. While the content of Ca in inclusions would continuously increase and finally, the inclusions would be with a liquid or semi-liquid layer or $CaO–Al_2O_3$ because of accumulated CaO in this surface layer. The evolution of spinel into calcium magnesia aluminate can be schematically shown in Fig. 5.30.

According to the study of Park [3] logarithm value of the diffusion coefficient for Ca was about -8.6 m^2/s, while the logarithm value of the diffusion coefficient for Al in calcium aluminate was about -10.4 m^2/s. It can be seen that the diffusion coefficient of Ca in calcium aluminate inclusions is much larger than that of Al. It has been known from above discussion that diffusion coefficient of Al in spinel is much larger than that of Mg. Therefore, diffusion coefficient of Mg in spinel should be much smaller than that of Al and Ca. Therefore, it can be known that diffusion rate of Mg in spinel should be the rate limitation step during the evolution of inclusions.

5.4 Summary

Based on the above analysis on the changes of inclusions in steel under refined by slag A with different refining time about 30, 60, 90 and 180 min, important recognition can be drawn. Calcium magnesia aluminate inclusions were with spinel as the core while a layer of $CaO–Al_2O_3$ surrounded outside. Moreover, the spinel core can be divided into two types. One type was featured by high MgO content and was written as $MgO–Al_2O_3$. While the other type was with high Al_2O_3 content and was

O because of accumulated CaO in this surface layer. The evolution of spinel into calcium

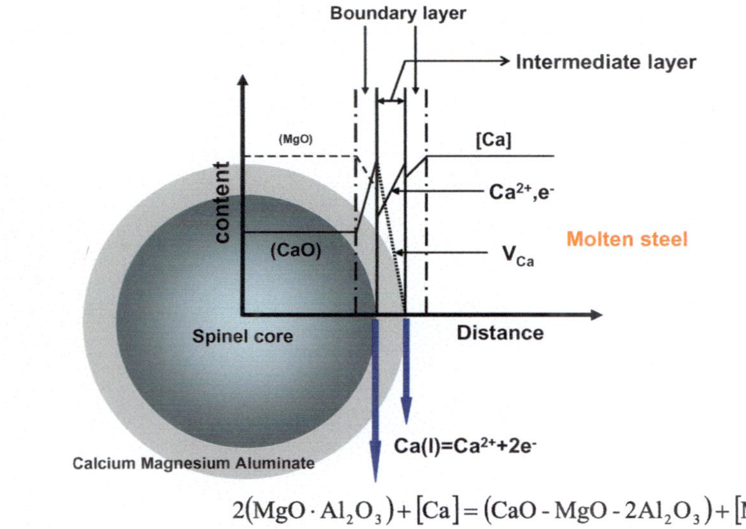

Fig. 5.30 Schematical description of spinel into calcium magnesia aluminate, reprinted from a previous paper of present authors on ISIJ International, 2010, vol. 50, pp. 96–104, with permission from ISIJ

written as Al_2O_3–MgO. By comparison, the latter occupied a larger fraction in the observed inclusions. During the slag-steel recation, inclusions experienced continous evolutions from Al_2O_3 to spinel and finally into low melting point calcium magnesia aluminate. During the transformation of inclusions, diffusion of Mg in inclusions was the rate limitation step.

References

1. Ye, G., Josson, P., Lund, T.: Thermodynamics and kinetics of the modification of Al_2O_3. Inclusions. ISIJ International **36**(Supplement), S105–S108 (1996)
2. Parra, R., Allibert, M.: Metal-slag reaction through a solid interfacial layer. Can. Metall. Q. **38**(1), 11–21 (1999)
3. Park, J.H., Kim, D.S., Lee, S.B.: Inclusion control of ferritic stainless steel by aluminum deoxidation and calcium treatment. Metallurgical and Material Transactions B **36B**, 67–73 (2005)
4. West, A.R.: Solid State Chemistry and its Applications, 4–6. Fudan University Press, Shanghai (1989) (Translated into Chinese by Su, M.Z., Xie, G.Y., Shen, P.W. et al.)
5. Sung, Y.M., Ahn, J.W.: Kinetic analysis of spinel formation at the interface of CaO-Al_2O_3 fibre-reinforced Al 7075 composites. J. Mater. Sci. **34**, 5127–5132 (1999)
6. Zayan, M.H., Jamjoom, O.M., Eazik, N.A.: High-temperature oxidation of Al-Mg alloys. Oxide Metal **34**(4), 323–333 (1990)
7. Kang, Y.J., Li, F., Morita, K., et al.: Mechanism study on the formation of liquid calcium aluminate inclusion from $MgAl_2O_4$ spinel. Steel Res. Int. **77**(11), 785–792 (2006)

Chapter 6
Influence of Refractory Material on the Formation Low-Melting-Point Inclusions

Abstract At present, MgO-based slag making materials and refractory are widely used in the industrial practice of steelmaking. Hence, effects of MgO-based refractory materials on the formation and evolution of inclusions in Al-deoxidized steel refined by high basicity high alumina slag should be evaluated. Considering that, present work was conducted, in which the ZrO_2 crucibles were used in the slag-steel reaction experiments. The obtained results showed that the choice of crucible importantly influenced the evolution of inclusions. In the ZrO_2 crucible experiments, low-melting-point inclusions can be formed within shorter time because the intermediate spinel inclusions can be avoided.

Control of non-metallic inclusions have been a very hot topic and intensively studied in recent years. Despite of studies on the "oxide metallurgy" technology focusing on utilizing inclusions with proper size distribution and chemistry to help refine microstructure of steel, inclusions are usually recognized to be harmful to the in-service behaviors of steel. Besides, inclusions, either in large size or with high melting point, are especially negative to the improvement of productivity. One of the typical problems caused by high-melting-temperature inclusions is the so-called nozzle clogging, which is initiated by the agglomeration of those inclusions on the inner wall of submerged entry nozzle (SEN) of continuous casting [1]. The clogging would result in a worse flow field of molten steel in casting mold. When cloggings were washed down by the pouring liquid steel into the casting mold, surface defects and fatigue breakage etc. would occur on final products. To solve these problems, as usual recognition, steelmakers should not only improve the cleanliness of molten steel but also need to target low melting temperature inclusions in steel.

Al is a popular deoxidizer for its strong affinity to oxygen. However, the produced high melting point alumina (Al_2O_3) inclusions tend to agglomerate into large-sized clusters and frequently observed to initiate SEN clogging [2]. Besides, high melting

A previous publication of present authors, reprinted by permission from springer nature, Metallurgical and Materials Transactions B, Formation of Low-Melting-Point Inclusions in Al-Deoxidized Steel Refined by High-Basicity Calcium Aluminate Slag in ZrO_2 Crucible Experiments, Min Jiang, Xinhua Wang and Jongjin Pak, copyright 2014, pp. 1248–1259.

M. Jiang and X. Wang, *Slag-Steel Reaction and Control of Inclusions in Al Deoxidized Special Steel*, Engineering Materials, https://doi.org/10.1007/978-981-19-3463-6_6

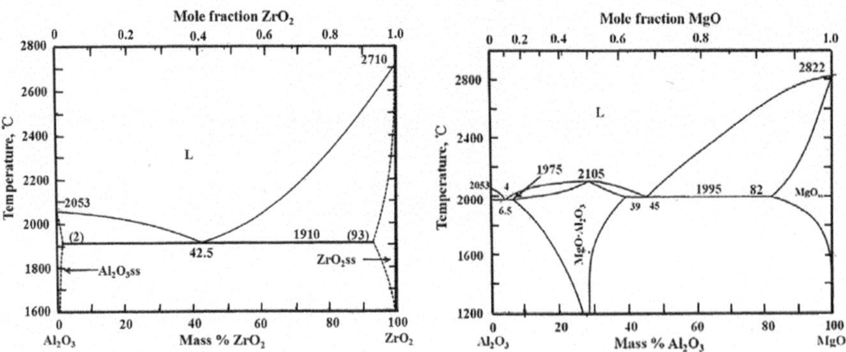

Fig. 6.1 Binary phase diagrams of Al_2O_3–ZrO_2 system and Al_2O_3–MgO system [18]

$MgO{\cdot}Al_2O_3$ (spinel) inclusion is another big trouble for Al killed steel. Actually, spinel can be easily formed because MgO-based refractory materials are widely used in steelmaking nowadays. As a result, spinel inclusions have been intensively studied in recent years [3–8]. Both of these two kinds of high-melting-point inclusions were very detrimental. For example, except nozzle cloggings, spinel inclusions are known as the main reason for rejected surface quality of stainless steel and decreased fatigue life span of bearing steel, as had been systematically reviewed by Park and Todoroki [9].

Therefore, up to now, there have been many papers focused on how to produce low-melting-point inclusions in Al deoxidized steel. Calcium treatment has been well known as a favorable operation in steelmaking process to modify Al_2O_3 into liquid calcium aluminates. It was also reported to be helpful to change $MgO{\cdot}Al_2O_3$ into low-melting-temperature CaO–MgO–Al_2O_3 inclusions [10–13]. Although Sakat mentioned that modification of $MgO{\cdot}Al_2O_3$ into MgO by Mg treatment can prevent the clogging in MgO–C nozzles, Mg treatment is a hard operation in industrial practice [14]. Proper choice of refining slag system was also proved to be an alternative useful method to target low-melting-point inclusions. Todoroki et al. reported that Al_2O_3 inclusions in Al deoxidized 304 stainless steel would be initially changed into $MgO{\cdot}Al_2O_3$ and finally to lower-melting-point CaO–MgO–Al_2O_3 during the refining by CaO–MgO–Al_2O_3–CaF_2 slag [15]. The present authors also discussed the evolutions of inclusions from $Al_2O_3 \rightarrow MgO{\cdot}Al_2O_3 \rightarrow$ low-melting-point CaO–MgO–Al_2O_3 during the refining by the high basicity calcium aluminate slag during the MgO crucible experiments [16, 17]. Referring to the MgO–Al_2O_3 binary phase diagram in Slag Atlas, it can be read that $MgO{\cdot}Al_2O_3$ was a very stable phase to be modified, as given in Fig. 6.1 [18]. Therefore, the use of MgO-based refractory material, which caused the formation of $MgO{\cdot}Al_2O_3$ during the evolution of Al_2O_3 to CaO–MgO–Al_2O_3, retarded the formation of liquid inclusion. ZrO_2 is another important refractory in steelmaking. By contrast, no intermediate compound would be formed between ZrO_2 and Al_2O_3 at steelmaking temperature, as can be seen from ZrO_2–Al_2O_3 phase diagram in Fig. 6.1. As a result, modification of Al_2O_3 to low melting point calcium aluminates can be expected to be easier when ZrO_2 refractory is used. However, the effect of ZrO_2 refractory on the formation of low melting point inclusions has not been fully investigated yet. Hence, it was one of the interests of present authors.

The above technologies are with regard to the control of inclusions to prevent nozzle clogging. As another approach, free-clogging nozzles have been developed by a number of engineers. In 1990s, much attention has been paid to ZrO_2-based refractory materials such as ZrO_2–CaO-graphite nozzle instead of Al_2O_3-based and MgO-based nozzles in the steelmaking industry. For example, Ogibayashi et al. studied the buildup mechanism of Al_2O_3 on submerged nozzle in continuous casting. They pointed out that CaO–ZrO_2-graphite nozzle can be free of clogging because a fine oxide layer composed of CaO–ZrO_2–Al_2O_3 liquid phase can be formed on the inner wall of the nozzle, which can be washed down by molten steel flow. As a result, the build-up of alumina inclusions was avoided [19]. Tsujino et al. studied the deposition of metal and inclusions in ZrO_2–CaO-graphite nozzle for steel with different Al contents and cleanliness [20]. Murakami et al. reassessed CaO–Al_2O_3–ZrO_2 phase diagram and found the optimal composition for tundish nozzle was CaO-saturated CaO·ZrO_2 [21]. These studies indicated the positive role of using ZrO_2 materials in solving the problems of high-melting-point inclusions in steel. And the work of Ogibayashi et al. implied that inclusions of CaO–Al_2O_3–ZrO_2 system should be formed by the reaction of alumina inclusions with the ZrO_2–CaO-graphite nozzle, which can be washed down the molten steel flow. Hence, CaO–Al_2O_3–ZrO_2 system inclusions can be expected after the molten steel passes through the nozzle. However, the primary chemistry of this inclusion system has not been so far studied in detail. In other words, the effect of ZrO_2 on the formation of inclusions with low melting points in Al deoxidized steel has rarely been studied.

Under such backgrounds, experiments were carried out to study the formation of low-melting-point inclusions during ZrO_2 crucible experiments, in which the low alloy steel melts were deoxidized by aluminum along with the high basicity calcium aluminate slag refining (very low level of SiO_2). Formation and evolution mechanism of inclusions were discussed and compared with the results obtained in previous MgO crucible experiments by the authors.

6.1 Laboratorial Experiments

6.1.1 Experimental Procedures

Slag-steel reaction experiments were carried out in a vertical electrical resistance furnace with $MoSi_2$ as heating bars. The chemical composition of steel material used in the experiment is given in Table 6.1. During each of the experiments, 100 g steel, 50 g slag and a proper amount of Fe–Al alloy (with Al about 40%) were charged into a ZrO_2 crucible (with inner diameter about 30 mm) and set in the even

Table 6.1 Chemical composition of steel material (mass%)

	C	Si	Mn	Cr
Content	0.58	1.90	0.85	0.19

temperature zone of the furnace. The targeted soluble aluminum content was about 0.03–0.04%. The furnace was heated to (1873 ± 2) K under the PID controlling mode. High purity argon gas was continuously introduced into the reaction tube during the experiments. 4 experiments with different holding times of 30, 60, 90, and 180 min were carried out separately at 1873 K. After each experiment, the crucible was taken out of the furnace and quenched in water. The obtained steel and slag samples were carefully prepared for the chemical analysis and the inclusion characterization. In the experiments, the used slag material was the mixture of chemical reagent CaO, SiO_2 and Al_2O_3 powder.

6.1.2 Chemical Analysis and Characterization of Inclusions

The obtained steel samples and slag samples were prepared for chemical analysis. While the chemical compositions of slag samples were measured by using an X-ray fluorescence spectrometer. Total oxygen contents in the steel samples were determined by the fusion and the infrared absorption methods. The contents of [Zr], [Al], and [Ca] in steel samples were analyzed using the ICP-AES method.

Species were also cut from steel samples and were ground and mirror-polished for inclusion detection. Hundreds of inclusions in each steel sample were randomly chosen and analyzed using an automatic SEM–EDS machine named the ASPEX explorer. Detailed descriptions of this instrument have been given in a previous paper by one of the present authors [22]. As a result, characteristics of inclusions such as morphology, size, and chemistry could be obtained.

6.2 Cleanness and Non-metallic Inclusions in Steel

6.2.1 Steel Composition and Slag Composition

Chemical compositions of both steel and slag samples are given in Table 6.2. Total oxygen (T.O) content in steel decreased from 0.0014% at 30 min to 0.0010% at 60 min, 0.0011% at 90 min, and 0.0009% at 180 min. Contents of [Al], [Ca], and [Zr] were in the range of 0.029%–0.036%, 0.0007%–0.0015%, and 0.0016%–0.0045%, respectively. Slags were featured with high basicity about 6.30–6.84 and high alumina content about 38.11%–43.82% with CaO/Al_2O_3 mass ratios about 1.0. ZrO_2 in slags increased continuously with the rise of reaction time, from 1.94% at 30 min to 4.49%

Table 6.2 Chemical compositions of steel samples and slag samples (mass %, mass ppm)

Exp (min)	Chemical compositions of steel							Chemical compositions of slag					
	C	Si	Mn	Al	Ca	Zr	T.O	CaO	Al_2O_3	SiO_2	FeO	MnO	ZrO_2
30	0.45	1.65	0.76	290	10	16	14	45.08	43.82	6.59	0.25	0.087	1.94
60	0.48	1.74	0.76	300	7	28	10	43.65	42.44	6.32	0.11	0.056	4.49
90	0.49	1.76	0.80	340	15	26	11	39.86	40.35	6.32	0.15	0.074	9.70
180	0.48	1.71	0.76	360	11	45	9	37.13	38.11	5.61	0.10	0.083	14.18

Note Cr in steel melts was 0.19 mass%, and units of Al, Ca, Zr and T.O contents are mass ppm

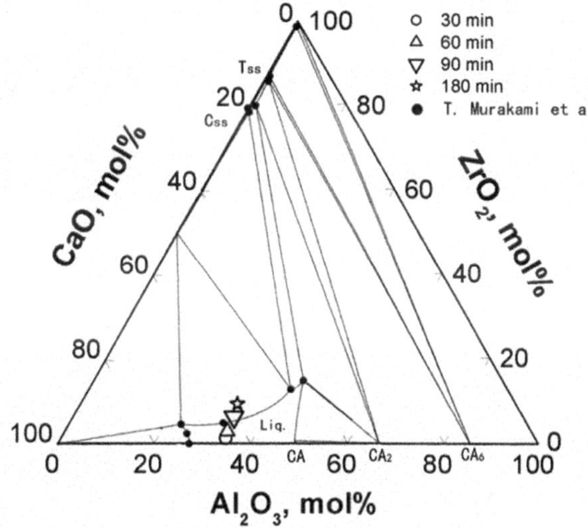

Fig. 6.2 Change of slag compositions with steel/slag reaction time, reprinted with experimental data points of present authors, with permission from springer nature, Metallurgical and Materials Transactions B, Phase diagram for the system $CaO–Al_2O_3–ZrO_2$, Murakami T, Fukuyama H, Kishida T, SUSA M and Nagata K, copyright 2000

at 60 min, 9.70% at 90 min and 14.18% at 180 min. As slags contained only a small amount of silica (about 5.61%–6.59%), therefore, compositions of the slags could be projected into the $CaO–Al_2O_3–ZrO_2$ ternary phase diagram reassessed by Murakami et al. as given in Fig. 6.2 [21]. It can be seen that slag compositions for the durations of 30, 60, and 90 min were in the liquid region, whereas, the slag composition moved out of this region and entered the coexistence region of solid and liquid phases at 180 min because of the rise of ZrO_2 content.

6.2.2 *Observation of Inclusions*

(1) Types and Morphology

The obtained results indicated that inclusions in steel were in spherical or near spherical shapes and with sizes of about 2–5 μm, as shown in Fig. 6.3.

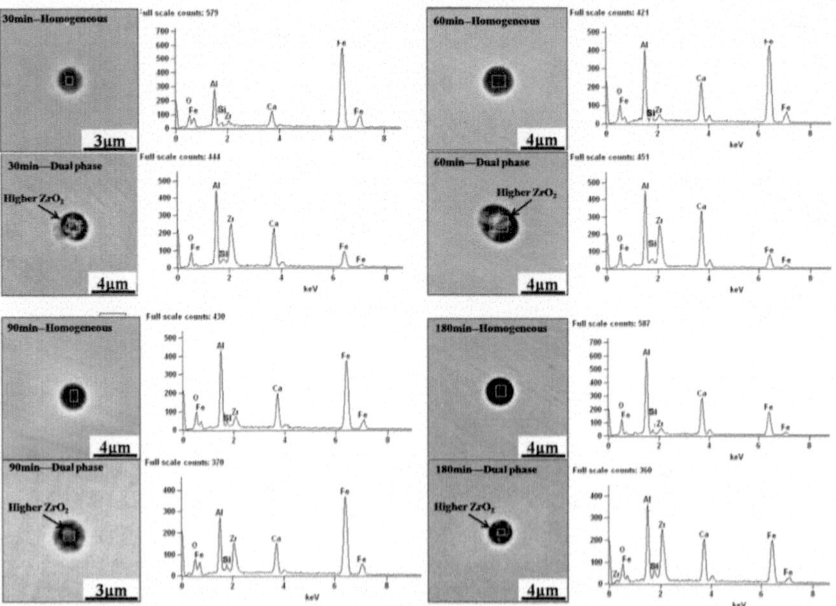

Fig. 6.3 Typical inclusions in steel samples at different reaction times (Fe is matrix)

From the EDS spectra, it can be seen that inclusions were mainly composed of CaO, Al_2O_3, and ZrO_2, with a very low level of SiO_2. And they can be classified into two types according to the height of the Zr peaks of the EDS spectra. The first kind contained a very low level of ZrO_2 with almost perfect spherical shape. The other type was impressed by much higher ZrO_2 and dual-phased structures, as can be seen in the pictures by the dark colored and grey-colored parts. Hence, images of the first kind of inclusions were labeled "30 min-Homogeneous", "60 min-Homogeneous", "90 min-Homogeneous" and "180 min-Homogeneous". The other type of inclusions was classified as "30 min-Dual phase", "60 min-Dual phase", "90 min-Dual phase", and "180 min-Dual phase".

SEM mappings of inclusions at different reaction times were given as Fig. 6.4, in which each image showed distribution of one certain element by the colored dots. And more dots meant larger amount of this element regardless the white or black colored background. As can be seen, CaO, Al_2O_3, and ZrO_2 homogeneously distributed in the first kind of $CaO–Al_2O_3–ZrO_2$ system inclusions. It was noticeable a much higher level of ZrO_2 existed and was locally concentrated inside the second type of inclusions like islands, as can be seen in the dual-phased image. Enrichment of ZrO_2 in the second type of inclusions can be clearly seen in the samples of 30, 60 and 90 min reaction, a typical line scanning of which was shown in Fig. 6.5. It can be read that a much higher level of ZrO_2 existed in the left half of this inclusion, which corresponded to the bright grey part of the inclusion. By contrast, the region enriched

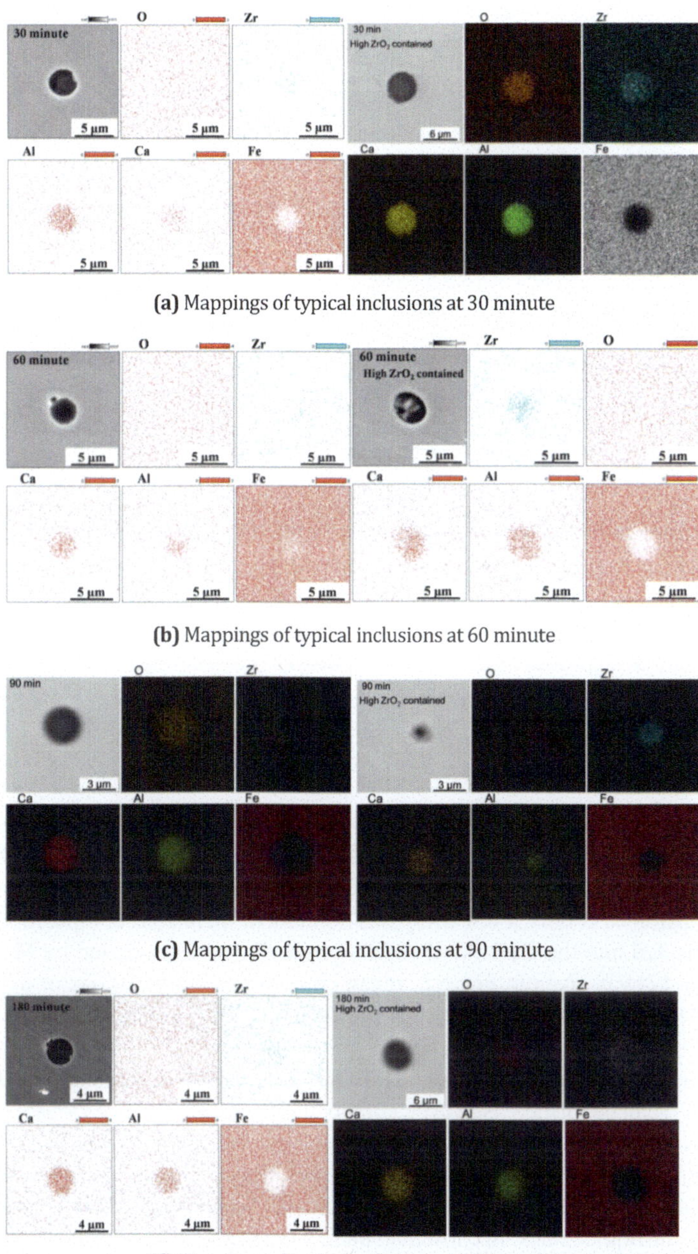

(a) Mappings of typical inclusions at 30 minute

(b) Mappings of typical inclusions at 60 minute

(c) Mappings of typical inclusions at 90 minute

(d) Mappings of typical inclusions at 180 minute

Fig. 6.4 SEM-mapping of $CaO–Al_2O_3–ZrO_2$ inclusions at different times

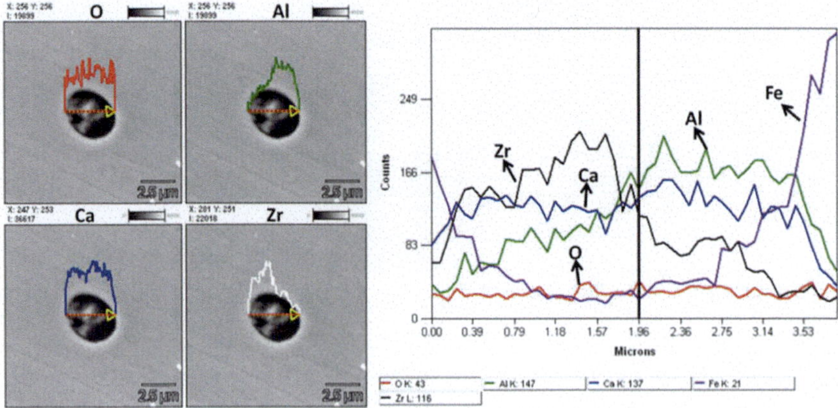

Fig. 6.5 SEM line scanning of the second type $CaO–Al_2O_3–ZrO_2$ inclusion

with Al was on the other half of the inclusion, while CaO homogeneously distributed in the whole inclusion. This phenomenon can be explained as following. According to the phase diagrams of $CaO–Al_2O_3$, $CaO–ZrO_2$ and $Al_2O_3–ZrO_2$, [18] no compounds would be formed between Al_2O_3 and ZrO_2. However, CaO can combine with both Al_2O_3 and ZrO_2 to produce calcium aluminates and $CaO·ZrO_2$. As a result, CaO was uniformly distributed in inclusion while segregation of Al_2O_3 and ZrO_2 occurred as the observatory.

(2) Composition Distribution

Weight percentages of CaO, SiO_2, Al_2O_3, and ZrO_2 in inclusions were estimated from the atomic percentages of Ca, Si, Al, and Zr obtained by the EDS method, supposing that a stoichiometric amount of oxygen was contained in the inclusions. As only a few SiO_2 was contained, inclusions compositions were projected into the $CaO–Al_2O_3–ZrO_2$ phase diagram reassessed by Murakami et al. [21] as given in Fig. 6.6. It can be seen that compositions of inclusions widely scattered in the phase diagram at 30 min. Only a small part of them were in the liquid region enlightened by grey color but many of the others were in high melting temperature regions where $CaO·6Al_2O_3$ (CA_6) and $CaO·2Al_2O_3$ (CA_2) can be formed. However, compositions of inclusions obviously concentrated after 60 min' reaction, mainly locating in the liquid or semi-liquid regions in grey color. After 90 min' and 180 min' slag-steel reaction, distribution of inclusion chemistry became more uniform and concentrated in the liquid phase or semi-liquid zones. In particular, there were some inclusions located near the corner of ZrO_2 or near the narrow area between T_{SS} (Tetragonal ZrO_2 solid solution) and C_{SS} all the time, which probably explained the observation of the above mentioned high ZrO_2 content inclusions. It was also noticed that the number of high ZrO_2 content inclusions at 60, 90 and 180 min were obviously more than that at 30 min. This can be attributed to the increase of ZrO_2 in

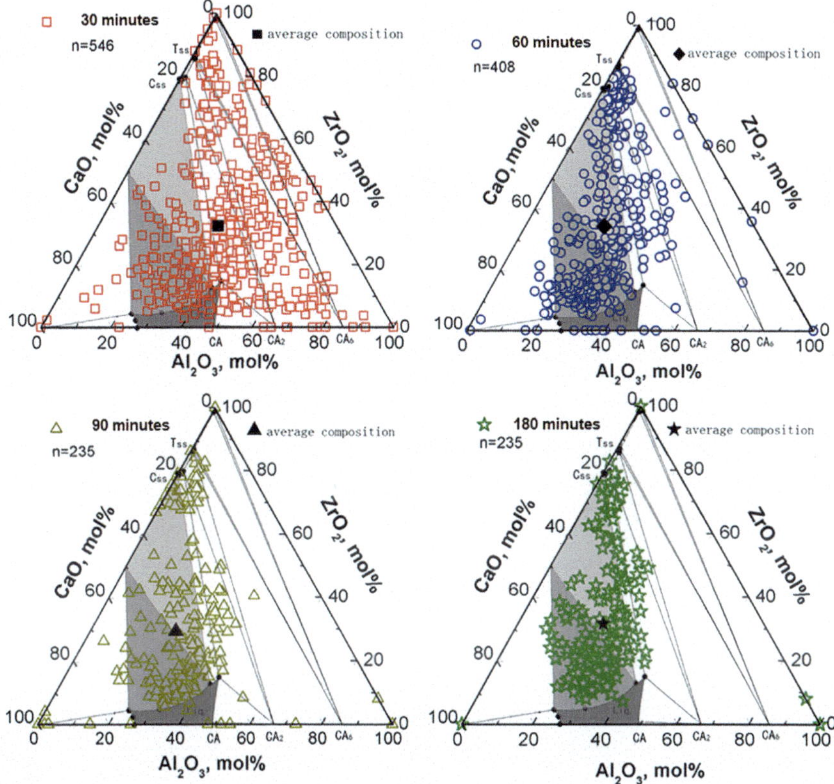

Fig. 6.6 Inclusion composition distribution in CaO–Al_2O_3–ZrO_2 ternary system

slag with the rise of reaction time because of the crucible erosion by the slag, which resulted in the increase of ZrO_2 activity in slag and in turn promoted the increase of ZrO_2 content in inclusions.

During the characterization of inclusions, every inclusion was chosen randomly and the SEM–EDS measurements were carried out automatically, which may partly contribute to the scattering of inclusion composition in. Therefore, it should be meaningful to take the average values to evaluate the inclusion chemistry at different times, as shown in Fig. 6.6 by solid data points. It can be seen that average composition of inclusion at 30 min was in the high melting point region, with ZrO_2 about 35%–40% and CaO/Al_2O_3 mass ratio about 1.0. While the average composition of inclusions moved into the hatched low melting temperature regions at 60, 90 and 180 min. It can be seen that the amount of ZrO_2 in the average composition was approximately the same at 30, 60, 90 and 180 min, while much higher CaO/Al_2O_3 mass ratios about 1.7–1.8 were obtained at 60, 90, 180 min. As a result, inclusion average composition entered into the shadowed low melting point regions of the phase diagram.

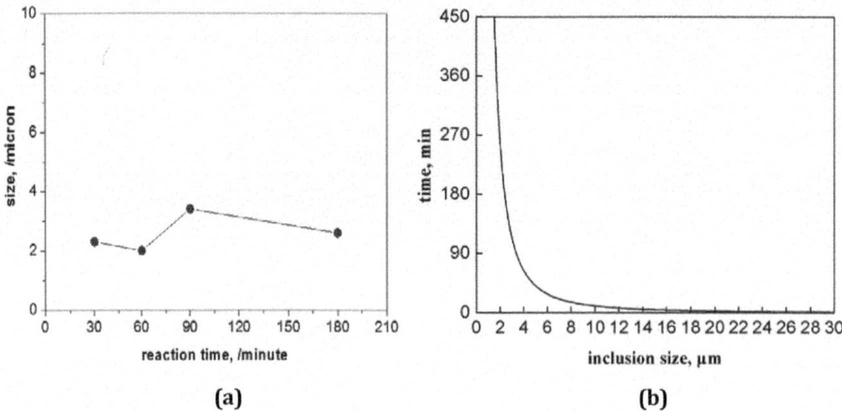

Fig. 6.7 Change of inclusion size with time: **a** Variation of average size with time in steel samples. **b** Plotting of inclusion size against flotation time estimated by Stokes Equation

Changes of inclusion sizes with the slag-steel reaction time were studied and the relation of inclusion size against time was indicated in Fig. 6.7. It can be seen from Fig. 6.7a that average size of inclusions was about 2–3 μm in all steel samples, although it showed a small increase at 90 min compared with the others. As no stirring was used to the melts during the experiments, flotation behavior of inclusion can be considered to follow the Stokes Equation given in Eq. (6.1), where, V is the velocity of inclusion particle, ρ_m is density of steel at 1873 K and taken as 7000 kg/m^3, ρ_{in} is the estimated average density of inclusion at 1873 K and taken as 3500 kg/m^3, u is the viscosity of steel at 1873 and taken as 0.005 $Pa \cdot s$, g is the acceleration of gravity and taken as 9.80 $kg \cdot m/s^2$ and d_{in} is diameter of inclusion. During the experiments, inner diameter of the used ZrO_2 crucible was 0.03 m. As 100 g steel material was used, depth of steel melts was about 0.02 m. Therefore, the floatation time for inclusions with varied size from the bottom of steel melts to steel/slag interface can be plotted as Fig. 6.7b. According to this curve, for inclusions with an average size about 2–3 μm, they would tend to remain in steel because very long time was needed for them to be removed by floatation. As a result, it can be pointed out that composition concentration of inclusions when reaction time was longer than 30 min was not caused by the floatation of solid inclusions scattered in the high melting point region, but because of the chemistry evolution of inclusion during steel/inclusion chemical reactions.

$$V = \frac{(\rho_m - \rho_{in})g \cdot d_{in}^2}{18u} \tag{6.1}$$

6.2.3 Formation Thermodynamics of $CaO–Al_2O_3–ZrO_2$ Inclusions

As Al was used as the deoxidizer, dissolved oxygen content in the steel melt can be estimated using the Al-O equilibrium, as expressed by Eqs. (6.2)–(6.5) [23].

$$(Al_2O_3)_{slag} = 2[Al] + 3[O] \tag{6.2}$$

$$\log K = 20.57 - 64{,}000/T \tag{6.3}$$

$$\lg f_{[i]} = \sum e_i^j [j\%] \tag{6.4}$$

$$\lg[\%O] = 1/3\left(\lg K - 2\lg f_{Al} - 2\lg[\%Al] - 3\lg f_O + \lg a_{Al_2O_3}\right) \tag{6.5}$$

where, K is the equilibrium constant, T is temperature in Kevin, f_i is the activity coefficient of component i in the 1 mass% standard state in liquid iron, and e_i^j is the activity interaction coefficient of element j on i in liquid iron. The standard state is taken as a pure solid as for a_i for component i in slag or inclusion.

During the experiments, slag compositions showed a continuous rise of ZrO_2 content, especially after 60 min' reaction. And slag composition moved out from the liquid region into the zone where $CaO{\cdot}ZrO_2$ can be formed, as given in Fig. 6.2 in section A. At the earlier stages, the liquid phases of the slags at 30 min and 60 min can be considered as composed of nearly $CaO–Al_2O_3$ binary system because of low level of ZrO_2. While slags at 90 min and 180 min were with ZrO_2 about 9.7% and 14.18%, respectively. Tanabe et al. studied the thermodynamics of $CaO–Al_2O_3$ and $CaO–Al_2O_3–ZrO_2$ systems [24]. Based on their results, the activity of Al_2O_3 in slag at 30 min and 60 min can be estimated as 0.412, while the activity of Al_2O_3 in slag at 90 min and 180 min can be estimated as 0.385. Activity coefficients can be calculated using Eq. (6.4), with the activity interaction coefficients listed in Table 6.3. Therefore, dissolved oxygen contents at different reaction times of 30, 60, 90, and 180 min after Al de-oxidation can be estimated about 0.0002%.

Table 6.3 Activity interaction coefficients between elements in liquid iron [23–27]

j i	Al	C	O	Ca	Mn	Si	Cr	Zr
Al	0.045	0.091	−6.6	−0.047	0.0065	0.0056	0.012	–
O	−3.9	−0.44	−0.2	−1040	−0.021	−0.131	−0.0459	−2.1
Ca	−0.072	−0.34	−9000	−0.002	−0.0156	−0.097	0.02	–
Zr	–	–	− 12	–	–	–	–	0

According to Al_2O_3–ZrO_2 and Al_2O_3–CaO binary phase diagrams, [18] there are no intermediate compounds between Al_2O_3 and ZrO_2. However, CaO can combine with ZrO_2 to form solid CaO·ZrO_2. In addition, CaO can also react with Al_2O_3 to produce various calcium aluminates. According to the composition distribution of the inclusions mentioned above, the possible liquid or t solid–liquid coexistence phases in inclusions may be 3CaO·Al_2O_3 (C_3A), 12CaO·7Al_2O_3 ($C_{12}A_7$), and CaO·Al_2O_3 (CA), etc. Berezhnoi and Kordyuk once mentioned that low melting point $Ca_7Al_6ZrO_{18}$ phase can be formed in the CaO–Al_2O_3–ZrO_2 system [28]. Espinosa and White agreed with that but they reported that this ternary compound was $Ca_{13}Al_{12}Zr_2O_{35}$ and that it would decompose into CaO·ZrO_2 and liquid phase by a peritectic reaction at 1813 ± 10 K [29]. It would be very hard to confirm whether this ternary compound formation and decomposition occurred in the inclusion during a high temperature experiment. So, here it was assumed that no ternary compound was formed during the formation of the CaO–Al_2O_3–ZrO_2 inclusion but binary compounds like solid CaO·ZrO_2 and solid and liquid coexisting xCaO·yAl_2O_3 could be formed. Therefore, during the formation of complex CaO–Al_2O_3–ZrO_2 inclusions, the related fundamental chemical reactions involved can be listed by following equations [23, 30–32].

$$(Al_2O_3)_{in} = 2[Al] + 3[O] \tag{6.6}$$

$$\log K = \log\frac{a_{[Al]}^2 \cdot a_{[O]}^3}{a_{Al_2O_3}} = 20.57 - 64{,}000/T \tag{6.7}$$

$$(CaO)_{in} = [Ca] + [O] \tag{6.8}$$

$$\log K = \log\frac{a_{[Ca]} \cdot a_{[O]}}{a_{CaO}} = -3.292 - 7220/T \tag{6.9}$$

$$(ZrO_2)_{in} = [Zr] + 2[O] \tag{6.10}$$

$$\log K = \log\frac{a_{[Zr]} \cdot a_{[O]}^2}{a_{ZrO_2}} = -57{,}000/T + 21.8 \tag{6.11}$$

$$(CaO) + (ZrO_2) = (CaO \cdot ZrO_2)_{in} \tag{6.12}$$

$$\Delta G^{\Theta} = -25{,}200(\pm 150) - 17.58(\pm 0.085)T \tag{6.13}$$

$$(CaO) + (Al_2O_3) = (CaO \cdot Al_2O_3)_{in} \tag{6.14}$$

$$\Delta G^{\Theta} = 59{,}413 - 59.413T \tag{6.15}$$

$$12(CaO) + 7(Al_2O_3) = (12CaO \cdot 7Al_2O_3)_{in} \quad (6.16)$$

$$\Delta G^{\Theta} = 617{,}977 - 612.119T \quad (6.17)$$

$$3(CaO) + (Al_2O_3) = (3CaO \cdot Al_2O_3)_{in} \quad (6.18)$$

$$\Delta G^{\Theta} = -21{,}757 - 29.288T \quad (6.19)$$

where, ΔG^{Θ} is the standard Gibbs free energy for formation in kJ/mol. The activities for CaO and Al_2O_3 used in the calculation for calcium aluminates were taken from the data of Fujisawa et al. [33]. And subscript "in" in the chemical reaction equations indicated the formation of inclusions.

Therefore, the stability phase diagram of CaO–CaO·ZrO_2–ZrO_2 and CaO–xCaO·yAl_2O_3–Al_2O_3 can be estimated, as given in Fig. 6.8. The former was calculated accounting for Ca and Zr contents under the condition of 0.03 mass% Al with soluble oxygen of 0.0002%. The latter was calculated accounting for Ca and Al contents with neglecting Zr content because liquid xCaO·yAl_2O_3 contains a negligibly small amount of ZrO_2 as can be read from the phase diagram of CaO–Al_2O_3–ZrO_2. It can be seen in Fig. 6.8a that CaO·ZrO_2 phase was stable over a wide range of Ca and Zr contents. This implied that a small amount of Ca stabilized CaO·ZrO_2 phase in the molten steel containing Zr, which was well consistent with the detections of ZrO_2 enriched inclusions and no formation of pure ZrO_2 phase in the inclusions. As a result, central parts of the inclusions were considered as CaO·ZrO_2 phase steadily produced during the experiments. The stability diagram of CaO–xCaO·yAl_2O_3–Al_2O_3 was shown as Fig. 6.8b. From the plot, it can be pointed out that the xCaO·yAl_2O_3 phase was CaO·Al_2O_3. In Fig. 6.9, the EDS result of a

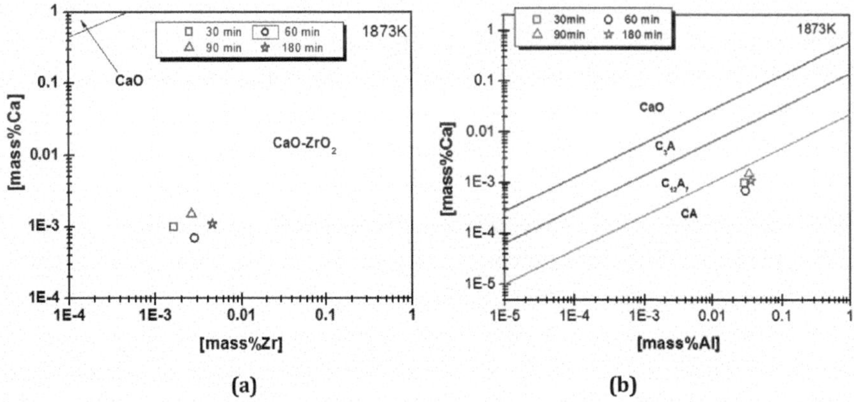

Fig. 6.8 Stability phase diagram of CaO–CaO·ZrO_2–ZrO_2 and CaO–xCaO·yAl_2O_3–Al_2O_3

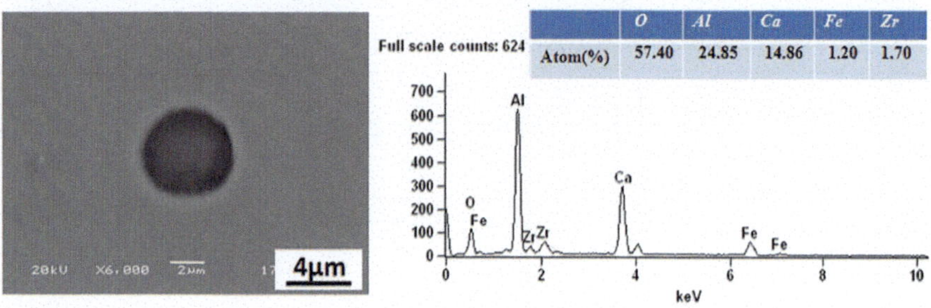

Fig. 6.9 Quantitative analysis result of typical homogeneous inclusions at 30 min

typical inclusion at 30 min was shown. Based on the atomic percentages of Ca and Al by EDS, molar ratio of Ca/Al in inclusion was about 1/2. As a stoichiometric amount of oxygen was supposed, CaO/Al_2O_3 molar ratio in the inclusion approximated to that in $CaO{\cdot}Al_2O_3$. Therefore, thermodynamic estimation agreed well with the experimental results at 30 min. In the steel samples of 60, 90 and 180 min, there were some inclusions with lower molar CaO/Al_2O_3 ratio but their numbers were limited. Referring to Fig. 6.8b, the average composition of inclusion at 60, 90 and 180 min should be similar to that at 30 min. However, the measured inclusion compositions were different from each other as have been shown in Fig. 6.6. So, differences existed between the experimental data and thermodynamic calculation. As usually recognized, it could be difficult to accurately determine the [Ca] content of very low level in steel at this moment. Therefore, analytical error might cause this inconsistency. Moreover, as inclusions were in the continuous evolution process during the experiments, their composition can be heterogeneous from the outer surface to the inner core. As a result, although the calculated calcium aluminate in inclusion was with high melting point, it was probable that liquid or semi-liquid calcium aluminate phase can be formed in the outer surface layer of the detected inclusions, which explained the spherical shapes of the inclusions even at 30 min.

Summarizing the above discussions, two types of $CaO–Al_2O_3–ZrO_2$ inclusions can be formed separately in liquid steel containing Ca, Zr and Al. Coexistence of solid $CaO{\cdot}ZrO_2$ phase and liquid $xCaO{\cdot}yAl_2O_3$ phase was quite possible according to the present analysis.

6.2.4 Effects of ZrO_2 Crucible on the Formation of Low Melting Point Inclusions

In previous works of present authors, MgO crucible experiments were carried out to target low melting temperature inclusions, in which the same steel material and slag system were used [16, 17]. It was found that reaction time affected significantly the type, composition, and morphology of inclusions. And inclusions experienced

Table 6.4 Comparison of cleanliness and inclusions in ZrO_2 and MgO crucible experiments

Present work						
Crucible	Cleanliness	Types of inclusions	30 min	60 min	90 min	180 min
ZrO_2	T.O: 0.0010% (average, > 60 min)	Homogeneous $CaO–Al_2O_3–ZrO_2$				
		High ZrO_2 contained $CaO–Al_2O_3–ZrO_2$			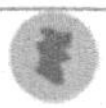	
Previous work by present authors [16]						
Crucible	Cleanliness	Types of inclusions	30 min	60 min	90 min	180 min
MgO	T.O: 0.0009% (average, > 60 min)	$MgO·Al_2O_3$				
		MgO-based				
		$CaO–MgO–Al_2O_3$				

a continuous evolution from $Al_2O_3 \rightarrow MgO·Al_2O_3 \rightarrow CaO–MgO–Al_2O_3$. Hereby, cleanliness and the characteristics of inclusions at varied reaction times in ZrO_2 crucible experiments and MgO crucible experiments were briefly compared, which can be clearly read from Table 6.4. In MgO crucible experiments, angular-shaped $MgO·Al_2O_3$ and MgO-based inclusions in solid state were formed in steel within 30 min. By contrast, all inclusions were already in spherical shape and composed of complex $CaO–Al_2O_3–ZrO_2$ system after 30 min reaction when ZrO_2 crucibles were used. Moreover, in MgO crucible experiments, although a small part of inclusions was with spherical shape, their composition located outside of the low melting point region. Liquid or semi-liquid inclusions were formed only after 90 min' reaction in MgO crucible experiments. However, liquid or semi-liquid inclusions can be largely produced in much shorter time in ZrO_2 crucible experiments, within 60 min. Cleanliness was compared by total oxygen (T.O) content. It was found that average T.O in the obtained steel samples approximated to each other for ZrO_2 and MgO experiments, about 0.0010% and 0.0009% respectively, showing no large differences.

As a result, it was considered that utilization of ZrO_2 crucibles accelerated the formation of liquid inclusions in a shorter time without deteriorating the cleanliness of the steel. It would be very helpful to minimize nozzle clogging in Al killed steel. For one hand, build-up of inclusions inside the nozzle can be effectively avoided. On the other hand, even the formed $CaO–Al_2O_3–ZrO_2$ layer inside the $CaO–ZrO_2$–graphite nozzle was washed down by molten steel flow into the casting mold as $CaO–Al_2O_3–ZrO_2$ system inclusions, which had been mentioned in the work of Ogibayashi et al. their negative effects can be minimized by the adjustment of steel compositions to change them into low melting point inclusions.

In the end, based on the obtained results, related mechanisms for the formation and evolution of inclusions in Al killed steel refined by high basicity calcium aluminate slag in ZrO_2 crucible experiments were explained as follows:

(1) After Al deoxidization, a lot of Al_2O_3 can be produced. Because of the complex chemical reactions among slag/steel/crucible/inclusions, the content of ZrO_2 in inclusions increased.
(2) During the experiments, Ca can be reduced from slag by Al in molten steel because FeO + MnO content in slag was of very low level, resulting in the rise of [Ca] in molten steel and the increase of CaO content in inclusions.
(3) According to the Al_2O_3–ZrO_2 binary phase diagram, no intermediate compounds would be produced between Al_2O_3 and ZrO_2 [18]. By contrast, in the MgO–Al_2O_3 binary system, Al_2O_3 and MgO react with each other to produce high melting point MgO·Al_2O_3, which was with a very stable lattice structure.
(4) As a result, during the reaction of steel/inclusion, it would be easier for [Ca] in molten steel to react and modify the solid solution of Al_2O_3–ZrO_2 than the condensed MgO·Al_2O_3 into produce liquid or semi-liquid phases in inclusions. Hence, low melting point inclusions can be targeted within shorter time when ZrO_2 crucible was used.

Further study was steel quite necessary in the future to make the evolution kinetics of inclusions clearer.

6.3 Summary

ZrO_2 crucible experiments were carried out in the lab to study the formation of low melting point of inclusions in steel, during which aluminum deoxidization and a high basicity calcium aluminate refining slag (CaO/SiO_2: 6%–8%, Al_2O_3 40%–45%) were used. 4 experiments were done with different slag/steel reaction time (30, 60, 90, and 180 min). It was found that inclusions were mainly composed of CaO–Al_2O_3–ZrO_2 with very limited SiO_2, in spherical morphology and with sizes mainly less than 5 μm. They can be classified into two types according to the ZrO_2 content. The first type contained much lower ZrO_2, whereas a much higher level of ZrO_2 was detected in the other kind. An evolution of inclusions with the reaction time was observed and studied. The obtained results indicated that chemical compositions of inclusions were widely scattered in the CaO–Al_2O_3–ZrO_2 phase diagram after 30 min reaction. However, inclusions composition became much more uniform and concentrated in low melting temperature regions at 60, 90, and 180 min, which would be favorable to prevent nozzle clogging and fatigue problems. Compared with the authors' previous results obtained in MgO crucible experiments, it was found that low melting temperature inclusions can be targeted in shorter time in ZrO_2 crucible experiments, without any degradation of cleanliness.

References

1. Singh, S.N.: Mechanism of alumina buildup in tundish nozzles during continuous casting of aluminum-killed steels. Metall. and Mater. Trans. B. **5**(10), 2165–2178 (1974)
2. Tozawa, H., Kato, Y., Sorimachi, K., et al.: Agglomeration and flotation of alumina clusters in molten steel. ISIJ Int. **39**(5), 426–434 (1999)
3. Park, J.H.: Thermodynamic investigation on the formation of inclusions containing $MgAl_2O_4$ spinel during 16Cr-14Ni austenitic stainless steel manufacturing processes. Mater. Sci. Eng., A **472**, 43–51 (2008)
4. Jiang, M., Wang, X.H., Chen, B., et al.: Formation of $MgO{\cdot}Al_2O_3$ inclusions in high strength alloyed structural steel refined by CaO-SiO_2-Al_2O_3-MgO slag. ISIJ Int. **48**(7), 885–890 (2008)
5. Seo, W.G., Han, W.H., Kim, J.S., et al.: Deoxidation equilibria among Mg, Al and O in liquid iron in the presence of $MgO{\cdot}Al_2O_3$ spinel. ISIJ Int. **43**(2), 201–208 (2003)
6. Nishi, T., Shinme, K.: Formation of spinel inclusions in molten stainless steel under Al deoxidation with slags. Tetsu-to-Hagané **84**(12), 837–843 (1998)
7. Itoh, H., Hino, M., Ban-Ya, S.: Thermodynamics on the formation of non-metallic inclusions of spinel ($MgO{\cdot}Al_2O_3$) in liquid steel. Tetsu-to-Hagané **84**, 85–90 (1998)
8. Kim, J.W., Kim, S.K., Kim, D.S., et al.: Formation mechanism of Ca-Si-Al-Mg-Ti-O inclusions in type 304 stainless Steel. ISIJ Int. **36**(1), S140–S143 (1996)
9. Park, J.H., Todoroki, H.: Control of $MgO{\cdot}Al_2O_3$ spinel inclusions in stainless steels. ISIJ Int. **50**(10), 1333–1346 (2010)
10. Seo, C.W., Kim, S.H., Jo, S.K., et al.: Modification and minimization of spinel ($Al_2O_3{\cdot}xMgO$) inclusions formed in Ti-added steel melts. Metall. and Mater. Trans. B. **41**(4), 790–797 (2010)
11. Verma, N., Pistorius, P.C., Fruehan, R.J., et al.: Transient inclusion evolution during modification of alumina inclusions by calcium in liquid steel: part II. Results and discussion. Metallurgical and Materials Transactions B **42**(4), 720–729 (2011)
12. Ito, Y.I., Nara, S., Kato, Y., et al.: Shape control of alumina inclusions by double calcium addition treatment. Tetsu-To-Hagane **93**(5), 355–361 (2007)
13. Higuchi, Y., Mumata, M., Fukagawa, S., et al.: Inclusion modification by calcium treatment. ISIJ Int. **36**(1), S151–S154 (1996)
14. Sakata, K.: Technology for production of austenite type clean stainless steel. ISIJ Int. **46**(12), 1795–1799 (2006)
15. Todorki, H., Mizuno, K.: Variation of inclusion composition in 304 stainless steel deoxidized with aluminum. Iron and Steelmaker Transaction **61**, 60–68 (2003)
16. Jiang, M., Wang, X.H., Chen, B., et al.: Laboratory study on evolution mechanisms of non-metallic inclusions in high strength alloyed steel refined by high basicity slag. ISIJ Int. **50**(1), 95–104 (2010)
17. Jiang, M., Wang, X.H., Wang, W.J.: Control of non-metallic inclusions by slag-metal reactions for high strength alloying steels. Steel Res. Int. **81**(9), 759–765 (2010)
18. VDEh: Slag Atlas, 2nd edn., pp. 39, 49, Verlag Sgahleisen GebH, Dusseldorf (1995)
19. Ogibayashi, S., Uchimura, M., Maruki, Y. et al.: Steelmaking Conference Proceedings, vol. 75, pp. 337–344. Iron & Steel Society, Toronto (1992)
20. Tsujino, R., Tanaka, A., Imamura, A., et al.: Mechanism of deposition of inclusions and metal on ZrO_2-CaO-C immersion nozzle in continuous casting. Tetsu-to-hagané **80**(10), 765–770 (1994)
21. Murakami, T., Fukuyama, H., Nagata, K., et al.: Phase diagram for the system CaO-Al_2O_3-ZrO_2. Metall. and Mater. Trans. B. **31**(1), 25–33 (2000)
22. Deng, X.X., Jiang, M., Wang, X.H.: Mechanisms of inclusion evolution and intra-granular acicular ferrite formation in steels containing rare earth elements. Acta Metallurgica Sinica. (English Letter) **25**, 241–248 (2012)
23. Hino, M., Itoh, H.: Recommended values of equilibrium constants for the reactions in steelmaking. Japan Society for the Promotion of Science **19**, 10 (1984)

24. Tanabe, J., Seki, I., Nagata, K.: Relationship between the aluminum and oxygen and sulfur partitions for molten iron and a $CaO–Al_2O_3–ZrO_2$ slag in equilibrium. ISIJ Int. **46**(2), 169–173 (2006)
25. Ohta, H., Suito, H.: Activities in $CaO-MgO-Al_2O_3$ slags and deoxidation equilibria of Al, Mg, and Ca. ISIJ Int. **36**(8), 983–990 (1996)
26. Sigworth, G.K., Elliott, J.F.: The thermodynamics of liquid dilute iron alloys. Metal Science **8**(3), 298 (1974)
27. Suito, H., Inoue, R.: Thermodynamics on control of inclusions composition in ultraclean steels. ISIJ Int. **36**(5), 528–536 (1996)
28. Berezhnor, A.S., Kordyuk, R.A.: Dokl. Akad. Nauk USSR **10**, 1344 (1963)
29. Espinosa, J., White, J.: Compatibility relationships in the system $ZrO_2-Al_2O_3-CaO$. Bol. Soc. Esp. Ceram. Vidrio **12**(4), 237 (1973)
30. Itoh, H., Hino, M., Ban-ya, S.: Deoxidation equilibrium of calcium in liquid iron. Tetsu-to-Hagané **83**(11), 695–700 (1997)
31. Tanabe, J., Nagata, K.: Use of solid-electrolyte Galvanic cells to determine the activity of CaO in the $CaO-ZrO_2$ system and the standard Gibbs free energies of formation of $CaZrO_3$ from CaO and ZrO_2. Metall. and Mater. Trans. B. **27**(4), 658–662 (1996)
32. Rein, R.H., Chipman, J.: Activities in the liquid solution $SiO_2-CaO-MgO-Al_2O_3$ at 1600°C. Transactions of Metallurgical Society AIME **233**(2), 415–425 (1965)
33. Fujisawa, T., Yamaguchi, C., Sakao, H.: Inclusion modification in Al-killed steel. In: Proceeding of the Sixth International Iron and Steel Congress, vol. 1, p. 201. ISIJ, Nagoya (1990)

Chapter 7
Prevention of CaO-Contained Inclusions in Steel During the Basic Slag Refining

Abstract From the laboratorial experiment results in previous chapters, it can be known that spinel inclusions would be hard to avoid during the massive steelmaking production, because of the wide use of MgO-based slag making materials and refractory materials. For example, during the ladle furnace (LF) refining in industrial production, it would be difficult to avoid spinel inclusions because of the easy proceeding of slag-steel reactions. To decrease spinel inclusions, some specific reactions between slag and steel should be prohibited. Based on this consideration, the present authors carried out exploring studies in the laboratory. Moreover, a technical concept of targeting "MnS + oxide" was also proposed and attempted in the experiment, with the aim of utilizing deformable MnS to wrap the un-deformable oxides and relieve the bad effects of such hard particles in steel.

High quality special steels are widely used in industry as bearings, gears, springs and so on. Except high strength and toughness, excellent fatigue resistance property is very important for them. As reported in many previous studies, inclusions are often observed as the origins of fatigue defects [1, 2]. Therefore, lower total oxygen (T.O) content is very helpful for decreased number of inclusions. Unfortunately, inclusions cannot be completely removed from steel. So, it would be more desirable if inclusions are with smaller sizes and lower melting points, which would be much favorable to prevent the formation of voids, cracks, stress concentration at steel-inclusion interfaces. Moreover, in some cases, machinability is demanded for some low oxygen special steels to obtain finished cutting surface, improved cutting efficiency and longer cutting tool life. In this case, sulfur is often intentionally added into steel to promote the massive formation of MnS with controlled numbers, sizes and uniform dispersion.

At present, a common process routine of low oxygen special steel is featured by Al de-oxidation and basic slag refining. Kawakami of Sanyo Special Steel in Japan

A previous publication of present authors, reprinted by permission from springer nature, Metallurgical and Materials Transactions B, Laboratory Study on Prevention of CaO-Containing ASTM "D-Type" Inclusions in Al-Deoxidized Low Oxygen Steel Melts During Basic Slag Refining, Min Jiang, Xinhua Wang, Die Yang, Shao-Long LEI and Kun-Peng WANG, copyright 2015, pp. 2573–2583.

M. Jiang and X. Wang, *Slag-Steel Reaction and Control of Inclusions in Al Deoxidized Special Steel*, Engineering Materials, https://doi.org/10.1007/978-981-19-3463-6_7

pointed out that T.O in high quality Al-deoxidized bearing steel can be reduced to about 5 ppm when the slag was with basicity and alumina about 4–7 and 20%–30%, respectively [3]. However, the problem is that high melting point inclusions, such as calcium aluminates, can be easily formed and the larger ones are very difficult to be completely avoided. Such calcium contained inclusions were quite detrimental despite their limited numbers. Referring to the ASTM-E45-05 Standard, such inclusions are classified as "D-type" inclusion [4]. If the sizes of these inclusions exceeded 13 μm, they were also named as the "DS-type" inclusions in the GB/T-10561–2005 Standard of China [5]. Kawakami reported that these lime-containing inclusions were originated from intrinsic de-oxidation/oxidation reactions among slag-steel-inclusions [3]. Ohta et al. of Kobe Steel also studied the formation of D-type inclusions in low oxygen bearing steel by industrial tracing trial during LF refining, but they considered that such kind of inclusions was mainly formed by the slag entrapment during refining [6]. Nowadays, with the development and technical progress in clean steelmaking in China, T.O has been decreased to a much lower level compared with before. However, control of these CaO-containing D-type or DS-type inclusions was found to be very difficult. They were limited in number but extremely detrimental to the fatigue property of steel. Hence, control of them has gradually become a hot research topic during the production of low oxygen special steel. For example, Miao et al. of reported that control of [Al], [Mg], [Ca] contents in liquid steel and the composition optimization of refining slag are very important for the control of D-type inclusions [7]. Li et al. of Baosteel also pointed that D-type inclusions can be reduced by decreasing the refining slag basicity and they pointed out that RH refining can be helpful in reducing D-type inclusions in bearing steel [8]. These studies helpfully reduced the population of DS inclusions. However, D-type inclusions cannot be thoroughly prevented.

It can be read from previous literatures that formation mechanisms of these lime-contained D-type inclusion in low oxygen special steels are not consistent although have been intensively studied. As a result, they are still big troubles remained to be solved. Moreover, if sulfur is added into low oxygen special steel for improved machinability, situation would be more complex because of the possible formation of CaS. As it known, CaS prefers to precipitate on the calcium aluminates, as reported by Sridhar [9], which is also with higher melting temperature and poorer deformability. From this point of view, it would be much better if CaO-contained inclusions can be completely prevented during the basic slag refining of Al deoxidized special steel. Nevertheless, the way on how to effectively or even completely prevent the formation of them was rarely reported. Under these backgrounds, thermodynamic analysis on formation of inclusions was carried out and experiments were designed by present authors to explore the possible way for complete prevention of CaO-containing inclusions in Al deoxidized low oxygen steel during basic slag refining.

7.1 Considerations on Inclusion Formation Thermodynamics and Design of Experiments

Thermodynamic reaction equilibrium among slag-steel-inclusions was usually evaluated to interpret the control of inclusions during slag refining, which can be briefly explained as following: (1) to make as whole as possible inclusions concentrated in a target composition region, concentrations of [Ca], [Mg], [Al], [O] etc. in liquid steel should be adjusted by the control of slag-metal reactions; (2) chemistry of inclusions were controlled through the reactions between [Ca], [Mg], [Al], [O], etc. and the inclusions.

For example, Eq. (7.1) can occur during the secondary refining of steel, where [Me] represents the dissolved Ca, Mg, Al, etc. while [O] is the dissolved oxygen, and $(Me_xO_y)_{slag}$ represents the formed reaction product such as CaO, MgO, Al_2O_3 etc. in the slag.

$$x[\mathrm{Me}] + y[\mathrm{O}] = \left(\mathrm{Me}_x\mathrm{O}_y\right)_{\mathrm{slag}} \tag{7.1}$$

When Reaction (7.1) between the slag and steel reaches equilibrium, the Gibbs Free Energy change of this reaction can be written as Eq. (7.2). When the equilibrium was established, the Gibbs Free Energy should equal to zero. Therefore, Eq. (7.3) can be deduced.

$$\Delta G_{\mathrm{slag/steel}} = \Delta G^{\Theta}_{\mathrm{slag/steel}} + RT\ln\frac{\left(a_{\mathrm{Me}_x\mathrm{O}_y}\right)_{\mathrm{slag}}}{a^x_{[\mathrm{Me}]}\cdot a^y_{[\mathrm{O}]}} \tag{7.2}$$

$$\Delta G^{\Theta}_{\mathrm{slag/steel}} = -RT\ln\frac{\left(a_{\mathrm{Me}_x\mathrm{O}_y}\right)_{\mathrm{slag}}}{a^x_{[\mathrm{Me}]}\cdot a^y_{[\mathrm{O}]}} \tag{7.3}$$

By comparison, there would also be the chemical reactions between [Ca], [Mg], [Al] with the inclusions inside the steel melt, which can be written as Eq. (7.4). Where, [Me] represents the dissolved Ca, Mg, Al, etc.in liquid steel, while [O] stands for the dissolved oxygen in steel and $(Me_xO_y)_{in}$ represents CaO, MgO, Al_2O_3 etc.in inclusions, respectively. Similarly, with the chemical reactions equilibrated between the inclusions and liquid steel, Eq. (7.5) can be obtained.

$$x[\mathrm{Me}] + y[\mathrm{O}] = \left(\mathrm{Me}_x\mathrm{O}_y\right)_{\mathrm{in}} \tag{7.4}$$

$$\Delta G^{\Theta}_{\mathrm{in/steel}} = -RT\ln\frac{\left(a_{\mathrm{Me}_x\mathrm{O}_y}\right)_{\mathrm{in}}}{a^x_{[\mathrm{Me}]}\cdot a^y_{[\mathrm{O}]}} \tag{7.5}$$

If same activity standard is adopted for components in slag and inclusions, for instance the Raoultian standard, the standard free energy change of Reaction (7.1) and Reaction (7.4) should have same value. As a result, Eqs. (7.7) and (7.8) can be deduced.

$$\Delta G^{\ominus}_{\text{in/steel}} = \Delta G^{\ominus}_{\text{slag/steel}} \tag{7.6}$$

$$-RT\ln\frac{\left(a_{\text{Me}_x\text{O}_y}\right)_{\text{slag}}}{a^x_{[\text{Me}]}\cdot a^y_{[\text{O}]}} = -RT\ln\frac{\left(a_{\text{Me}_x\text{O}_y}\right)_{\text{in}}}{a^x_{[\text{Me}]}\cdot a^y_{[\text{O}]}} \tag{7.7}$$

$$\left(a_{\text{Me}_x\text{O}_y}\right)_{\text{in}} = \left(a_{\text{Me}_x\text{O}_y}\right)_{\text{slag}} \tag{7.8}$$

As a result, it can be seen that chemistry of inclusions would follow and approximate to the composition of the top slag with the approaching of chemical reaction equilibrium among slag/steel/inclusions, which theoretically explained why CaO be very easily contained in the inclusions in Al deoxidized low oxygen special steel after basic slag refining. Therefore, if CaO-containing D-type inclusions should be prevented, some countermeasures should be done to prevent reactions between slag and steel, which in turn affects the formation of inclusion in steel by chemical reactions between inclusion and steel melt.

During the basic slag refining of Al deoxidized steel, oxides like FeO, MnO, CaO and MgO in slag can be reduced by [Al] in steel melt, as expressed by chemical reactions (7.9), (7.10), (7.11) and (7.12). Theoretically, if (FeO + MnO) content in slag was of very low level, occurrence of reactions (7.11) and (7.12) would be much easier and there would be a rise of [Ca] and [Mg] content in liquid steel. Correspondingly, contents of CaO and MgO in inclusions would be increased. However, if (FeO + MnO) content or activity in slag of higher level, then reactions (7.11) and (7.12) would be suppressed or retarded. Hence, contents of [Ca] and [Mg] can expected to be lower, resulting in the decrease of CaO and MgO content in inclusions.

$$3(\text{FeO})_{\text{slag}} + 2[\text{Al}] = (\text{Al}_2\text{O}_3) + 3[\text{Fe}] \tag{7.9}$$

$$3(\text{MnO})_{\text{slag}} + 2[\text{Al}] = (\text{Al}_2\text{O}_3) + 3[\text{Fe}] \tag{7.10}$$

$$3(\text{CaO})_{\text{slag}} + 2[\text{Al}] = (\text{Al}_2\text{O}_3) + 3[\text{Ca}] \tag{7.11}$$

$$3(\text{MgO})_{\text{slag}} + 2[\text{Al}] = (\text{Al}_2\text{O}_3) + 3[\text{Mg}] \tag{7.12}$$

Based on the above analysis, several countermeasures and hypothesis were taken during the experiments which also needed to be verified by the experiments, including:

(1) Fe_2O_3 powder was intentionally added into the initial slag materials during the experiments to increase the initial FeO content in top slag. The most important aim, as mentioned above, is to prohibit or retard the reducing of CaO in slag by [Al] in liquid steel, which was expected to be effective to decrease the CaO content in inclusions.

(2) If CaO in inclusions can be effectively reduced, then the formed inclusions would possibly be solid alumina or spinel. In this case, intensive stirring would be favorable to promote the floatation of inclusions, resulting in the reduction in the number and size of the solid inclusions. Therefore, an vacuum induction furnace was thought to be better and was chosen for the experiments.
(3) If the formed inclusions after experiments were still CaO-contained, then it can be pointed out they were not formed by intrinsic reactions, and vice versa.
(4) During the experiments, the used slag volumes were similar to or less than that of ladle refining in industrial production.

7.2 Experimental Procedures

7.2.1 Deoxidization and Refining Experiments

The experiments were carried out in a vacuum induction furnace. In each experiment, 5 kg electrolytic iron bar with surface finished well by dry turning was charged into the MgO crucible. Initial compositions of the iron were given in Table 7.1. Then the chamber was vacuumed and followed by the refilling of high purity argon gas for three times before the electrolytic iron was heated. With the temperature reached about 1873 K, aluminum alloy was added for strong deoxidation. After Al deoxidization, high purity graphite, silicon, electrolytic Mn, FeS powder etc. were then added into the steel melt to adjust chemical compositions. Then, the prepared slag material was charged into the crucible, which was the mixture of CaO, SiO_2, Al_2O_3, CaF_2 and Fe_2O_3 powder. With the melt held for predetermined time (about 25 min), molten steel was sampled by a bucket shaped sampler for total oxygen analysis. The remained liquid steel in the crucible was tapped from the MgO crucible into the ingot mold made of casting iron, which was originally set inside the reaction chamber. And the steel samples were cut from the ingot for inclusions analysis. Unfortunately, slag samples were very hard to be collected after the experiments because the very limited volume.

Three groups of experiments (group I, group II and group III) were totally carried out, as shown in Table7.2. During the experiments, influence of initial oxygen contents in steel melts, Al_2O_3 contents in slag and slag volume etc. on the cleanliness and formation of inclusions were taken into account. In group I, no Al_2O_3 was contained in slag and the slag volumes were two times of that in group II and group III. For group II and group III, influence of initial oxygen contents in molten steel before Al de-oxidation were increased, during which initial oxygen in the melts in

Table 7.1 Chemical compositions of the used electrolytic iron (mass%)

Element	C	Si	Mn	P	S	Al	Ni	Cu	T.O
Content	0.003	0.007	0.09	0.008	0.006	0.01	0.01	0.01	0.0032

Table 7.2 Initial conditions of the used slag in the experiments

Group	Sample	Basicity	Al_2O_3	Fe_2O_3	CaF_2	Slag volume (g)
I	A	3	0	10%	10%	100
	B	3	0	10%	10%	100
II	C	3	5%	10%	10%	50
	D	3	10%	10%	10%	50
III	E	3	5%	10%	10%	50
	F	3	10%	10%	10%	50

group III was increased by melting the electrolytic iron in the air before deoxidation (special precautions were done to secure the safety). In group II and III, two experiments were carried out, respectively, to investigate effects of Al_2O_3 content (5 and 10%) in slag on the formation of inclusions. All the used basic slags were with basicity (weight percentage of CaO to SiO_2) about 3 and contained about 10% CaF_2.

7.2.2 Sample Preparation and Analysis

Steel samples were prepared for chemical analysis. Acid-soluble Al, Ca and Mg content in steel were analyzed by ICP-MS method. Total oxygen content in steel sample was determined by fusion and infrared absorption method. Unfortunately, slag was very hard to be collected after the experiment because of the tapping of liquid steel into the ingot mold and also very limited slag volume.

Steel samples for inclusions analysis were also prepared. Inclusions on the cross-sectioned plane of each steel sample were randomly analyzed by an automatic Scanning Electron Microscopy (with Energy Dispersive Spectrometer) named ASPEX PSEM explorer (ASPEX Co. Ltd., now FEI Co. Ltd.). During the analysis of inclusions, the automatic mode of this equipment was used because much larger area was observed than the manual mode. As a result, many inclusions can be detected and careful analysis on their size and chemistry can be carried out.

7.3 Results and Discussions

7.3.1 Chemical Compositions of Steel Samples

Chemical compositions of steel were list in Table 7.3. It is noted that dissolved aluminum and sulfur content in the melts varied in the scope of 0.0030%–0.2000% and 0.0026%–0.0270%, respectively. It can be read from the table that very high cleanliness of steel melts was obtained in all steel samples, with T.O contents in the

Table 7.3 Chemical compositions and cleanliness of steel (mass%)

Group	Sample	C	Si	Mn	Al	S	Mg	Ca	T.O
I	A	0.53	0.54	1.16	0.0045	0.0088	0.0004	0.0002	0.0007
	B	0.66	0.66	0.99	0.0074	0.0028	0.0005	0.0001	0.0010
II	C	1.0	0.07	0.54	0.0190	0.024	0.0005	0.0001	0.0005
	D	0.48	0.14	0.56	0.0200	0.0190	0.0004	0.0001	0.0010
III	E	0.62	0.54	1.47	0.2200	0.0195	0.0005	0.0002	0.0006
	F	0.32	0.44	0.94	0.2000	0.0440	0.0005	0.0002	0.0003

range of 0.0003–0.0011%. It can be seen from the table that T.O contents in the two experiments with high initial oxygen contents before deoxidization were higher than other experiments.

As it known, total oxygen content includes the dissolved oxygen and oxygen in oxide inclusions. During Al deoxidation, oxygen potential was controlled by Al–O reaction equilibrium, as can be expressed as Eqs. (7.13), (7.14) and (7.15).

$$2[\mathrm{Al}] + 3[\mathrm{O}] = (\mathrm{Al_2O_3}) \tag{7.13}$$

$$\log K_{13} = \frac{a_{\mathrm{Al_2O_3}}}{a_{[\mathrm{Al}]}^2 \cdot a_{[\mathrm{O}]}^3} \tag{7.14}$$

$$\log f_{[i]} = \sum \mathrm{e}_{[i]}^{[j]} \cdot [\%j] \tag{7.15}$$

where, a_i standards for the activity of component i in steel melt or slag, K is the equilibrium constant of Eq. (7.13), $f_{[i]}$ is the activity coefficient of element i in steel melt, e_i^j is the activity interaction coefficient of element j to i. Based on the Eqs. (7.13), (7.14) and (7.15), dissolved oxygen after Al deoxidization in all the three groups of experiments can be estimated to be about 1–2 ppm. For experiments in group III, much higher initial oxygen contents before deoxidization can be expected as the steel materials was melted in the air. As a result, more inclusions can be expected to be produced in the Al deoxidation. As the floatation time of inclusions were the same in all the experiments, more inclusions can be probably remained in the steel melts of group III, resulting higher total oxygen contents. So, it can be reasonably pointed out that higher total oxygen contents in the group III experiments was originated from large number of oxide inclusions remained in liquid steel. And it was indicated that lower initial oxygen content before deoxidation would be beneficial to improve the cleanliness, which also agrees the common sense.

7.3.2 Typical Inclusions in Steel Samples

Non-metallic inclusions in the steel samples were intensively detected for the detailed information of them. It was found that observed inclusions in steel were singular MnS in irregular shapes (occupying a most majority patio), singular oxides of MgO–Al_2O_3 or Al_2O_3, and complex dual-phased (MgO–Al_2O_3) + MnS or Al_2O_3 + MnS. In particular, CaO-contained inclusions were not inspected. The typical inclusions in all the steel samples were indicated in Figs. 7.1, 7.2, 7.3, 7.4, 7.5 and 7.6. As it can be seen, the singular MnS inclusions were with varied morphologies in different samples, which can affect the machinability of steel but was not the emphasis of

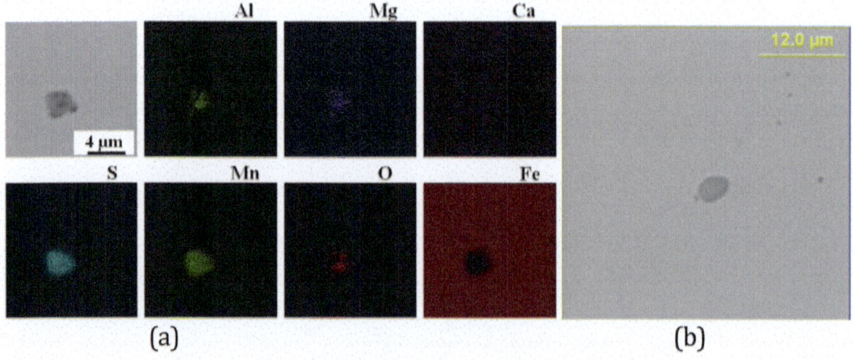

Fig. 7.1 Typical inclusions in steel sample A. **a** dual-phased (MgO–Al_2O_3) + MnS inclusion. **b** Singular MnS inclusion

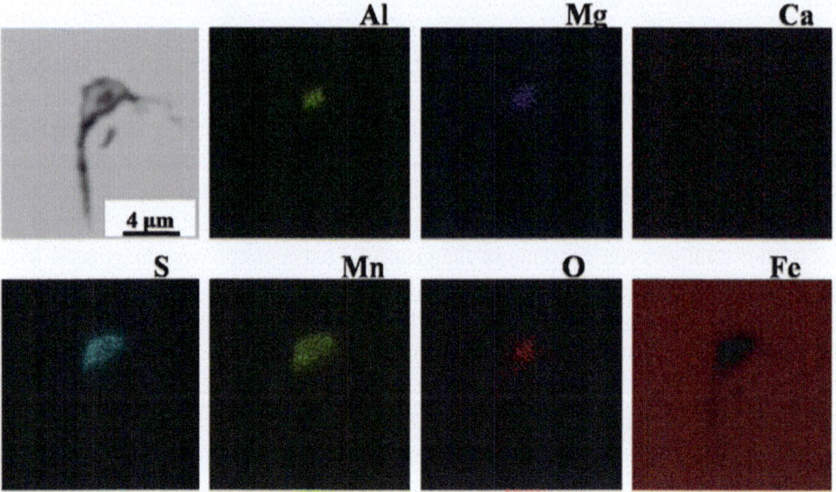

Fig. 7.2 Typical inclusions in Steel sample B: Dual-phased (MgO–Al_2O_3) + MnS inclusion

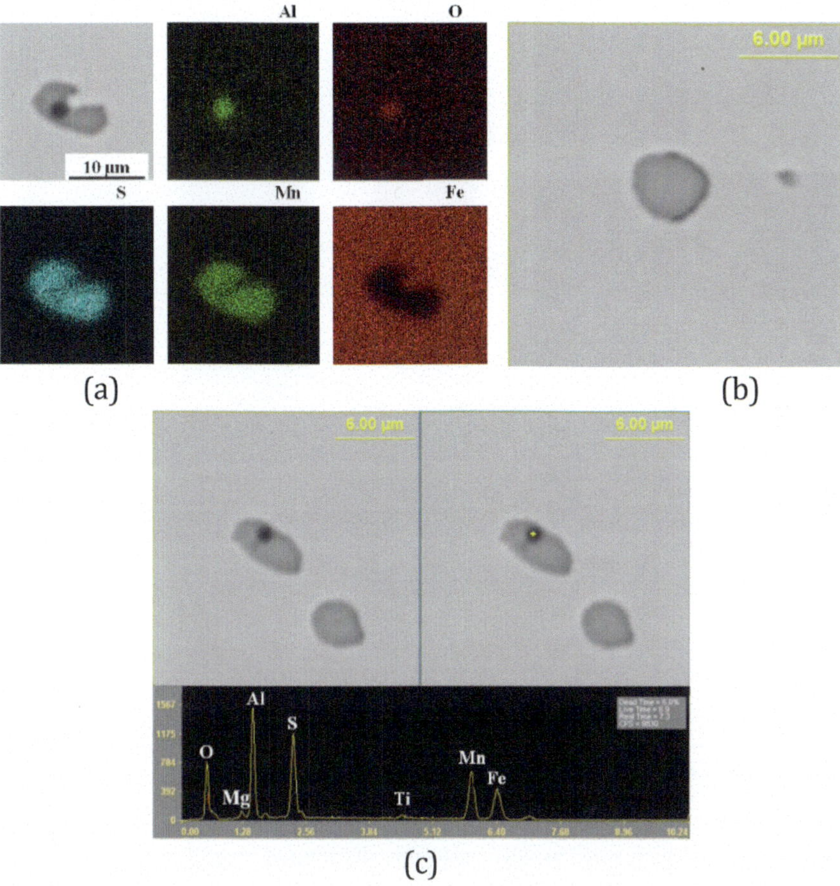

Fig. 7.3 Typcial inclusions in sample C. **a** Dual-phased (Al_2O_3 + MnS). **b** Singular MnS. **c** Dual-phased (MgO–Al_2O_3 + MnS) inclusion

present paper. What should be noted is that, SEM-mappings of the formed (MgO–Al_2O_3) + MnS and Al_2O_3 + MnS dual-phased complex inclusions were characterized by very tiny MgO–Al_2O_3 and Al_2O_3 (about 1–3 μm or even smaller in size) as the cores, edging in the relative much thicker MnS outer surface layers, as can be seen in Figs. 7.1, 7.2, 7.3, 7.4, 7.5 and 7.6. As it known, MnS inclusions can be deformed very well together with the steel matrix, which were with much better deformability compared with most of the oxide inclusion. It can be reasonable pointed out that this kind of complex inclusions should be very positive to avoid stress concentrations at inclusion/steel matrix interfaces and therefore effectively improve the fatigue resistance property of low oxygen special steel. It may be probably attributed to the differences in contact angles between MgO–Al_2O_3 and steel and between Al_2O_3 and steel. As has been mentioned in many previous works, wettability of Al_2O_3 in liquid steel were much poorer, resulting in a much stronger tendency of agglomeration than

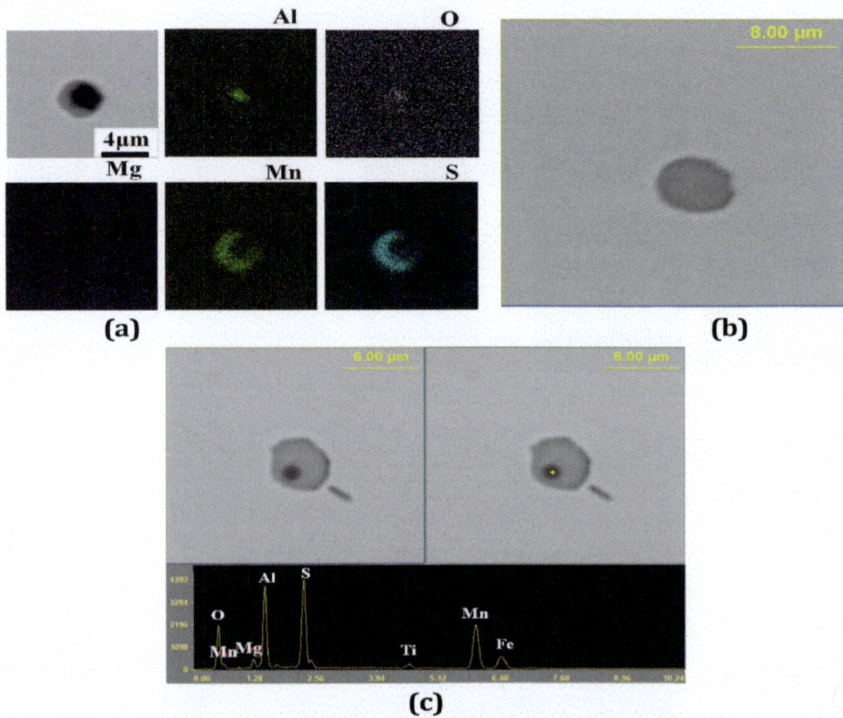

Fig7.4 Typical inclusions in sample D. **a** Dual-phased (Al_2O_3 + MnS). **b** Singular MnS. **c** Dual-phased ($MgO–Al_2O_3$ + MnS) inclusion

$MgO–ssAl_2O_3$, which were usually uniformly distributed in steel in very tiny sizes. Moreover, it is not hard to find that types of the inclusions were closely related to the [Al] and [S] contents in steel melts. In the three groups of experiments, [Al] contents in steel melts were gradually increased. As a result, the produced oxides were changed from $MgO–Al_2O_3$ inclusions in group I to Al_2O_3 inclusions in group II and group III. This is understandable because of there have been many works of the formation of $MgO–Al_2O_3$ spinel and Al_2O_3 in the past. Moreover, although there were still some $MgO–Al_2O_3$ oxides as the cores of the "oxide + MnS" complex inclusions in the experiments of group II and group III, but these inclusions occupied very limited percentages in populations and were featured with much lower level of MgO contents, resulting a difficulty in showing the existence of MgO by SEM-mapping method. As a result, existence of the very low contents of MgO was shown by the SEM–EDS spectra, as given in Figs. 7.3c, 7.4c, 7.5c and 7.6c. By comparison, formation of MnS inclusions was influenced by [S] contents in steel melts. For example, in the experiments of group I, [S] content in sample A was about 0.0088%, higher than that in sample B. Correspondingly, singular MnS inclusion can be frequently observed in sample A while not singular MnS inclusion was much

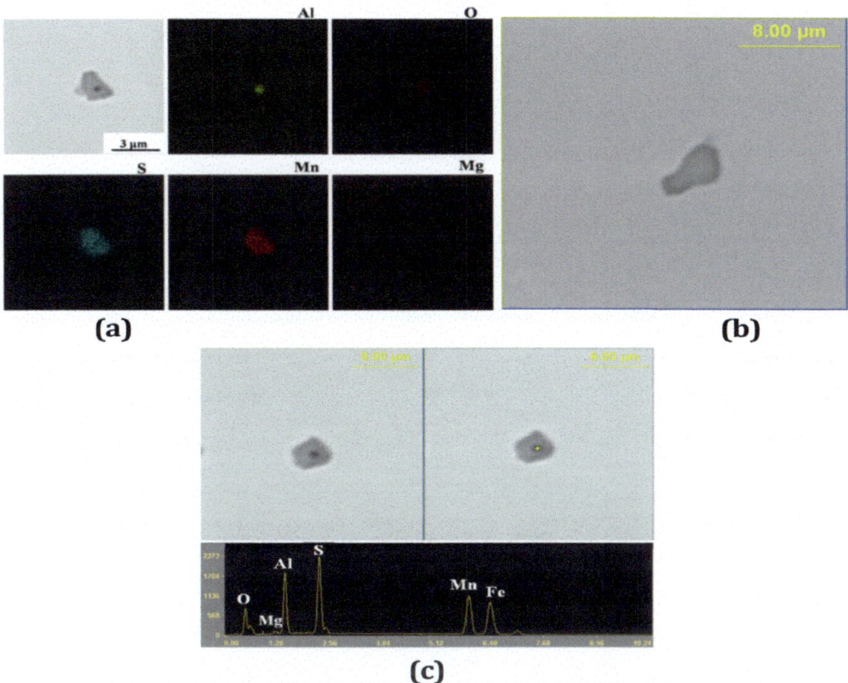

Fig. 7.5 Typical inclusions in sample E. **a** Dual-phased ($Al_2O_3 + MnS$) inclusion. **b** Singular MnS. **c** Dual-phased ($MgO–Al_2O_3 + MnS$) inclusion

less found in sample B. In other two groups of experiments with even higher [S] contents, singular MnS inclusions were much largely produced.

Size distributions of the inclusions in steel samples were also analyzed and shown in Fig. 7.7. As can be read, most majority of the inclusions were mainly with sizes in within 13 μm. There was only a very small part of inclusions exceeded 13 μm and it was noticed that such larger inclusions were all singular MnS inclusions, as given in Fig. 7.8. As it known, morphology and size of MnS was closely related to the chemical compositions of steel and also solidification conditions, which should be carefully controlled for resulfurized steel grades required better machinability. However, this is not focused in present paper. Moreover, as mentioned at the beginning of this section, it was noticed that not all of the oxide inclusions were wrapped by the MnS outer surface layers. There are still some singular tiny $MgO–Al_2O_3$ and Al_2O_3 inclusions existed in the steel samples but they occupied a very limited ratio. As can be seen in Fig. 7.9, over 90% percent of the $MgO–Al_2O_3$ or/and Al_2O_3 were wrapped by the MnS surface layers. Of the six heats of experiments, four of them with the ratios larger than 95% and even to 99%. And it was noticed that these unwrapped oxide inclusions showed no differences in the size distribution with other oxides that were wrapped by MnS layers. This difference may be initiated during the solidification of steel, which directly influenced the heterogenous nucleation behaviors of MnS

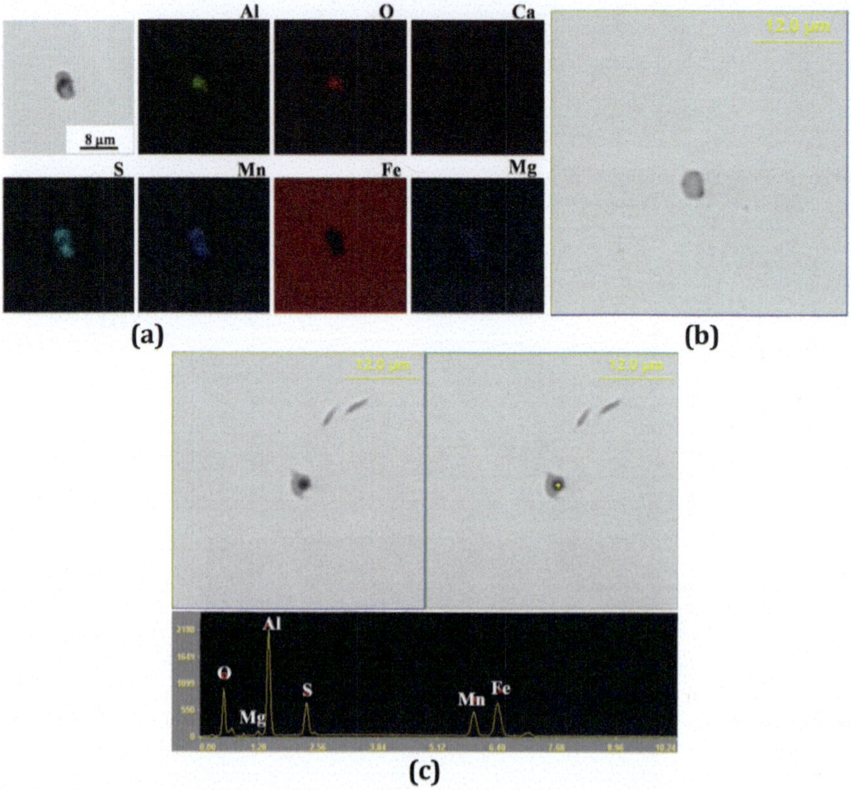

Fig. 7.6 Typical inclusions in sample F. **a** Dual-phased ($MgO–Al_2O_3 + MnS$). **b** Singular MnS. **c** Dual-phased ($MgO–Al_2O_3 + MnS$) inclusion

Sample	A	B	C	D	E	F
Number	3276	445	4552	3121	6808	5838
Area, /mm²	56.7	63.18	53.46	56.7	45.36	25.11

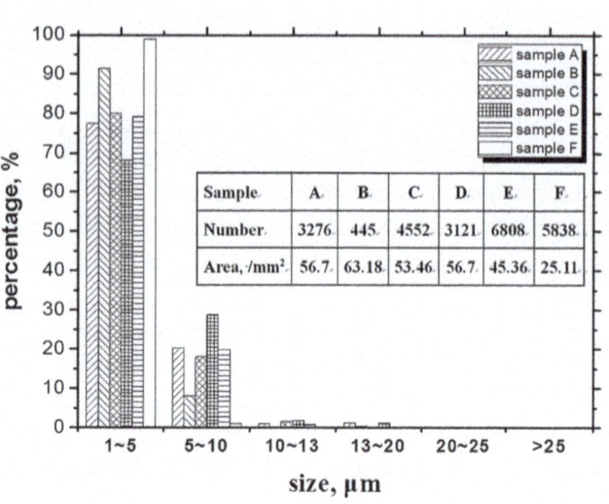

Fig. 7.7 Size distribution of the observed inclusions in the obtained steel samples

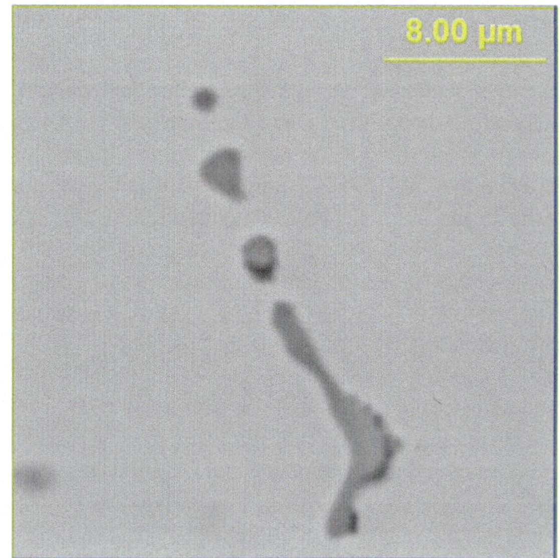

Fig. 7.8 Typical singular MnS with size over 13 μm

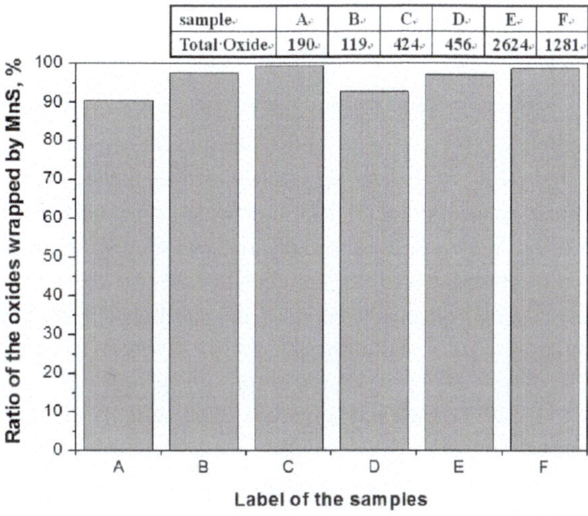

Fig. 7.9 Ratio of $MgO–Al_2O_3$ or/and Al_2O_3 wrapped by MnS surface layers

on oxides. However, this need further and detailed investigation in the future. Much more important, it can be seen that formation of CaO-contained D-type and DS-type inclusions were successfully prevented in the experiments.

7.3.3 Influences of Slag/Steel Reactions on Inclusions

As can be seen from the obtained results given above, CaO-contained inclusions were successfully prevented. There were many studies either thermodynamic calculation or experimental works on the formation of inclusions in Al deoxidized steel melts. As it known, MgO–Al_2O_3 spinel inclusion were always found to be formed very easily even [Al] and [Mg] contents were of very low level. And according to the classical work of Itoh et al. and recent typical works of Pistorius et al. and the systematical review of Park and Todoroki, if some [Ca] entered into the solution even about 0.0002%, the formed MgO–Al_2O_3 spinel can be changed into calcium aluminates, that is, CaO-contained inclusions [10–12]. In this study, the produced oxide inclusions were either MgO–Al_2O_3 or pure Al_2O_3 even basic refining slag was used and [Ca] in liquid steel was about 0.0002%.

The obtained results and the formation of inclusions can be understood by the comparison of industrial practice of ladle furnace (LF) refining of low oxygen special steel and the practice of ultra-low carbon steel by BOF-RH process. As it known, usually, ladle furnace refining (LF) is often used in the production of low oxygen special steel and this facility is especially common in China. To reduce the T.O in steel, basic slag with very low level of Fe_tO, for example less than 0.5%, is often recommended during the LF refining. Because of argon gas stirring at the bottom of the ladle and electric arc heating, conditions for slag/steel reactions would be good. Therefore, equilibrium can be much easier to be approached or approximate. And a definite rise of [Ca] can occur, attributing to the formation of CaO-contained inclusions. By contrast, in the production of ultra-low carbon steel, BOF is closely followed by RH decarburization during the production of ultra-low carbon steel. Initial Fe_tO content of RH slag was higher and commonly equals to that of BOF ending slag because of the slag carry-over in BOF tapping. After the decarburization, Fe_tO contents were even higher because of oxygen blowing. And also because of the limited refining time after the addition of Al in RH and higher Fe_tO in slag compared to LF, reducing of CaO in top slag by [Al] in liquid steel would be difficult. Therefore, the intrinsic inclusions in ultra-low carbon steel were mainly alumina while CaO-contained inclusions were seldom observed although the basic top slag is used.

As a result, formation of the oxide inclusions in present study was explained as following:

(1) After the addition of aluminum alloy for de-oxidation, many alumina inclusions were formed in steel melts. Because of the strong stirring in the induction furnace, there would be good flotation of the alumina inclusions.
(2) Because of the added Fe_2O_3 powder in the initial slag, there would be higher Fe_tO contents in slag. As a result, reducing of MgO or CaO from crucible or slag by the steel melts would be suppressed or retarded.
(3) However, with the proceeding of the reactions, for example between the steel and the MgO-based crucible and between the steel and CaO in slag, there were

definitely a rise of [Mg] and [Ca] in steel melt. As a result, formation of MgO–Al_2O_3 spinel can be thermodynamically possible. For steel melts with not so high level of [Al], existence of the formed MgO–Al_2O_3 spinel would be more stable.

(4) Because MgO–Al_2O_3 spinel can be formed much easier and earlier than the CaO-containing inclusions, which has been discussed in the authors' previous works [13, 14], a rise of CaO content in inclusions was successfully prevented. As a result, CaO-containing D-type inclusions were prevented.

Therefore, in the end, the authors want to cautiously point out that a refining process featured by BOF-RH can be possibly suitable for the production of Al deoxidized low oxygen special steel in order to prevent the formation of CaO-containing D-type inclusions or even larger sized DS-type inclusions. However, this needed to be verified in industrial practice in the future. Moreover, much further and detailed works are still quite needed to find out deeper and more accurate mechanisms on formation of CaO-containing D-type and DS-type inclusions. And effects of the slag compositions and slag volumes on the formation of inclusions should also be investigated further.

7.4 Summary

Laboratory experiments were carried out by a 5 kg scale vacuum induction furnace to investigate the prevention of CaO-containing D-type inclusions (ASTM standard E45-05) in Al deoxidized low oxygen steel. Different from ordinary concept that chemical reaction equilibrium should be promoted among slag/steel/inclusions for better control of inclusions during basic slag refining, chemical reaction between slag and steel was intentionally suppressed to decrease the reducing of CaO in slag by the steel melt with the aim to decrease the CaO contents in inclusions. The obtained results and conclusions can be summarized as following:

(1) Very high cleanliness was realized in the steel melts. T.O contents in all the steel samples were of very low level, in the range of 0.0003–0.0010%. More important, CaO-containing oxide inclusions were successfully prevented in all the experiments.

(2) The observed inclusions were singular MnS inclusions, dual-phase (MgO–Al_2O_3) + MnS or Al_2O_3 + MnS inclusions and a very small part of singular MgO–Al_2O_3 or Al_2O_3. And the dual-phased (MgO–Al_2O_3) + MnS or Al_2O_3 + MnS complex inclusions were characterized by very tiny MgO–Al_2O_3 or Al_2O_3 about 1–3 μm edged in the thicker MnS outer surface layers. As MnS is can be well deformed during hot rolling, such complex inclusions are thought to be positive to improve the fatigue resistance property of special steel.

(3) Types of the formed inclusions greatly depended on the [Al] and [S] contents in steel. For example, in steel samples with lower [Al] contents about 0.0045 and 0.0074%, the formed oxides were MgO–Al_2O_3. While in steel samples with

higher [Al] contents about 0.019% or even higher to about 0.22%, the inspected oxides were mainly pure Al_2O_3 and only very small part of $MgO–Al_2O_3$ were formed. By comparison, [S] in steel largely influenced the formation of MnS inclusions. In steel samples with [S] over 0.0088%, singular MnS inclusions can be largely observed.

(4) It was noticed that in steel samples with higher [Al] contents about 0.019 to 0.22%, not all the formed Al_2O_3 and $MgO–Al_2O_3$ were wrapped by the MnS layers. And this part of oxides showed no obvious differences in size distributions with the ones wrapped by MnS. It may be related to the behaviors of oxides and precipitation of MnS during the solidification of steel. However, more detailed investigations are quite needed in the future.

References

1. Sun, C., Lei, Z., Xie, J., et al.: Effects of inclusion size and stress ratio on fatigue strength for high-strength steels with fish-eye mode failure. Int. J. Fatigue **48**(3), 19–27 (2013)
2. Lei, Z., Hong, Y., Xie, J., et al.: Effects of inclusion size and location on very-high-cycle fatigue behavior for high strength steels. Mater. Sci. Eng., A **558**(12), 234–241 (2012)
3. Kawakami, K.: Generation mechanisms of non-metallic inclusions in high-cleanliness steel. Sanyo Technical Report **14**(1), 22–35 (2007)
4. ASTM E 45-5: Standard Test Methods for Determining the Inclusion Content of Steel (2005)
5. GB/T10561-2005-2005/ISO 4967. 1998(E): Steel-determination of content of nonmetallic inclusions-micrographic method using standards diagrams, general administration of quality supervision. Inspection and Quarantine of the People's Republic of China/Standardization Administration of the People's Republic of China (2005)
6. Ohta, H., Kimura, S., Mimura, T., et al.: Behavior of CaO containing inclusions during ladle refining of ultraclean bearing steel. Kobe Steel Engineering Report **61**(1), 98–101 (2011)
7. Miao, X.D., Yu, C.M., Shi, C.M., et al.: Formation and control of calcium-aluminates inclusions in bearing steel. J. Univ. Sci. Technol. Beijing **29**(8), 771–775 (2007)
8. Li, Z., Hu, J.H., Xu, M.H., et al.: Influence of refining slag composition on D type inclusion in bearing steel. In: Proceeding of Baosteel BAC 2006, Shanghai, pp. 411–413 (2006)
9. Wang, Y., Sridhar, S., Valdez, M.: Formation of CaS on Al_2O_3-CaO inclusions during solidification of steels. Metall. and Mater. Trans. B. **33**(4), 625–632 (2002)
10. Itoh, H., Hino, M., Ban-Ya, S.: Thermodynamics on the formation of spinel nonmetallic inclusion in liquid steel. Metall. and Mater. Trans. B. **28**(5), 953–956 (1997)
11. Verma, N., Pistorius, P.C., Fruehan, R.J., et al.: Calcium modification of spinel inclusions in aluminum-killed steel: reaction steps. Metall. and Mater. Trans. B. **43**(4), 830–840 (2012)
12. Park, J.H., Todoroki, H.: Control of $MgO{\cdot}Al_2O_3$ spinel inclusions in stainless steels. ISIJ Int. **50**(10), 1333–1346 (2010)
13. Jiang, M., Wang, X., Chen, B., et al.: Laboratory study on evolution mechanisms of non-metallic inclusions in high strength alloyed steel refined by high basicity slag. ISIJ Int. **50**(1), 95–104 (2010)
14. Jiang, M., Wang, X.H., Wang, W.J.: Control of non-metallic inclusions by slag-metal reactions for high strength alloying steels. Steel Res. Int. **81**(9), 759–765 (2010)

The manufacturer's authorised representative in the EU is Springer Nature Customer Service Centre GmbH, Europaplatz 3, 69115 Heidelberg, Germany. If you have any concerns regarding our products, please contact ProductSafety@springernature.com

Printed and bound by CPI Group (UK) Ltd, Croydon, CR0 4YY
15/07/2026
02167655-0001